SCIENCE AND ENGINEERING POLICY SERIES

General Editors The late Lord Jackson of Burnley
Sir Harrie Massey
Sir Frederick Dainton

Economic and Social Consequences of Nuclear Energy

Edited by Lord Sherfield

OXFORD UNIVERSITY PRESS 1972

Oxford University Press   *Ely House, London, W.I*

Glasgow	Delhi
New York	Bombay
Toronto	Calcutta
Melbourne	Madras
Wellington	Karachi
Cape Town	Lahore
Ibadan	Dacca
Nairobi	Kuala Lumpur
Dar es Salaam	Singapore
Lusaka	Hong Kong
Addis Ababa	Tokyo

Printed in Great Britain by
William Clowes & Sons Ltd.
London, Colchester and Beccles

Contents

Introduction
by Lord Sherfield   1

1   The science of the fission and fusion processes
    by Professor Robert Spence, C.B., F.R.S., University of
    Kent   7

2   The technological development of the uranium and
    'hydrogen' bombs
    by Air Chief Marshal Sir Denis Barnett, G.C.B., C.B.E.,
    D.F.C.   24

3   The nuclear power programme
    by Sir Stanley Brown, C.B.E., Central Electricity Generat-
    ing Board   34

4   The potentialities of the development of nuclear energy for
    purposes other than warfare and power
    by the late Dr. Hans Kronberger, C.B.E., F.R.S.   54

5   The moral aspects
    by The Right Revd. Robert C. Mortimer, Bishop of Exeter 69

6   The sociological consequences of nuclear energy
    by Lord Ritchie-Calder   76

Index   87

Introduction

By Lord Sherfield: sometime Chairman of the U.K. Atomic Energy Authority

The idea of this book was conceived by the late Lord Jackson of Burnley, who was an editor of this series until his untimely death, and he and I discussed its form and content together. Lord Jackson died before the contributions could be assembled, and as editor of this book I have therefore been deprived of his wise counsel in carrying through the project. Unhappily too, Dr. Kronberger died shortly after delivering his manuscript and before he could see his chapter through the press.

The contributors are all pre-eminent in their fields; they approach their subjects in many ways and write in very different styles, but the broad conclusions which emerge from their contributions are nevertheless harmonious.

There have been many books written about 'the Atom' and 'the Bomb', and they exhibit almost the whole gamut of possible judgements, prejudices, and emotions on the subject of nuclear energy. The year 1970 was the twenty-fifth anniversary of the destruction of Hiroshima and Nagasaki, and after a quarter of a century the political, economic, and sociological effects of those events were still reverberating round the world, and had only partly been absorbed, synthesized, and understood. This is perhaps the justification for another book on the subject.

The controversy about the Anglo-American decision to employ the weapon against Japan was given another airing in 1970. It has been increasingly suggested in recent years that the use of the bomb was not a military necessity, and that there were alternative courses of action which would have brought about a Japanese surrender without further heavy fighting and loss of life. As one who was closely involved in the events leading up to the decision, I have never accepted the view that an alternative method of instantly terminating the war existed in 1945, and it is not without interest that both Group Captain

Cheshire, with his personal experience of the operation, and Mr. Laurens van der Post, with his profound knowledge of Japanese psychology, wrote on the anniversary in support of this view.

But the argument will continue, and there is little doubt that if the atomic weapon shocked the Japanese into surrender (as General Marshall believed it did), it also shocked the people of the world into a state of apprehension about all uses of nuclear energy, peaceful as well as warlike. It has been a case of guilt by association. This has cloaked and somewhat vitiated the brilliant scientific and industrial achievement which the realization of the weapon has represented, and which is described by Sir Denis Barnett in Chapter 2.

This reaction was natural enough. As Professor Spence explains in Chapter 1, information about the fundamental scientific research needed for the development of this weapon was in print and publicly available before August 1939; indeed, this was the reason for supposing that the Germans would develop it. But the fact that the bomb was developed in complete secrecy, and that the secrecy was maintained after the event, heightened the popular fears which were naturally generated.

The beneficial applications of nuclear energy were clearly apprehended by the British scientists working on the military applications in 1940, and formed part of the post-war British nuclear programme from its start in 1946, but this was masked both by the primary military objective and the security aspects of the project. It was later further obscured by the fact that the path to military and to peaceful applications is the same for three-quarters of its length, and that weapon materials, electric power, and radioisotopes are all products of an essentially similar industrial process. A long and complex industrial route requiring heavy and expensive capital investment must be completed before the path diverges. Indeed, a power reactor is a source of plutonium and radioisotopes as well as electrical energy, though, of course, the conversion of the resultant nuclear material into weapons requires the most intricate engineering technology.

The secrecy surrounding the production and application of atomic energy was of great concern to scientists, some of whom were very uneasy about their part in bringing the nuclear weapon to fruition. In December 1953, partly owing to this and partly to more general political considerations, a new departure in policy was made by General Eisenhower, then President of the United States. By that date it was clear that the Soviet Union possessed all the knowledge required to make sophisticated weapons. It was also clear that the post-war efforts to bring about a system of international control of the manufacture and use of nuclear weapons had failed. The President concluded that the time had come to make public as much knowledge as possible about the peaceful uses of atomic energy.

His speech resulted in the convening of the so-called 'Atoms for Peace' conference at Geneva in August 1955. At this conference scientists from all the nuclear powers expounded the potentialities of the new source of energy and aroused the highest expectations in the public mind. The possession of a nuclear reactor, however modest, became a symbol of prestige for every nation, great or small, developed or undeveloped. About the same time a national power programme was introduced in several countries, led by the United Kingdom. The claims of the scientists were usually hedged around by cautionary words, but these were generally overlooked in the popularizing process. In the United Kingdom, in particular, a very large programme of nuclear power stations was undertaken. This was certainly justified at the time on the basis of the economic forecasts, both national and international, of the anticipated availability in Europe of conventional fuels in respect both of supply and foreign exchange and of the cost of producing power by conventional means.

Economic forecasting is prone to fallibility, and it is not surprising that the forecast on many aspects of the problem turned out to be wrong. As the nuclear stations were being designed and built, the proven world reserves of oil and gas increased, the exchange problems were reduced, and the cost of the latest types of conventionally fuelled power stations progressively fell.

All these events combined to produce in the first instance, a sense of rising expectation, even of euphoria, about the benefits of the new technology. The public was led to expect too much too quickly. The claims were not wrong in themselves, but the time-scale in which they could be realized was not fully spelt out and appreciated. Nuclear energy was oversold. In the same period, suggestions which had been advanced about the possibility of producing economic thermonuclear power from the fusion of the light elements proved to be premature.

There followed a period of disenchantment. The small reactors which so many countries had bought were expensive to run and were of little use except to train nuclear physicists and engineers. The nuclear power stations required heavy capital investment and, though they operated successfully, their economics became a question of bitter and often excited controversy. Sir Stanley Brown gives a very clear account in Chapter 3 of what is involved in estimating the comparative merits of different methods of producing electric power.

Nevertheless, after all the ups and downs, the balance sheet after a quarter of a century is favourable to nuclear energy.

On the military side, positive progress towards international control of nuclear armaments has been very slow, some would say non-existent. Yet the international agreements for the control of nuclear testing and for the non-proliferation of nuclear weapons are not negligible achievements. On the negative side, there is a strong presumption that the existence of nuclear armaments is the strongest safeguard against the outbreak of a major war. It is the essence of an ultimate weapon that it can never be used, but paradoxically, it seems to follow that the possessors of these weapons are handicapped in bringing their power to bear on less well-armed countries which are bent on fighting each other. The nuclear stalemate appears to be stable, especially if the two major nuclear powers can agree on an assessment of its nature and its merits. But, as the Bishop of Exeter so strongly emphasizes, the nuclear bomb ought not to be regarded as 'just another weapon'. Its existence increases and extends the moral responsibility of the leaders of the nations which possess it.

The most important economic consequence of nuclear energy is that it has provided the world with another source of fuel which in one of its forms is self-renewing. The introduction of nuclear power into the power systems of industrial countries is taking place slowly and with the circumspection required with a new fuel which could cause a major accident. But nuclear power has now passed through the phase in which it was introduced into many countries for reasons of national prestige or to learn a new technology, and is being treated on its economic merits in relation to other fuels. It is making headway everywhere, and the introduction of the fast breeder reactor to supplement the thermal reactors now in use is likely to accelerate the use of nuclear power in the next decade.

Much is written about the hazards of nuclear power. In fact there has never been an accident to a civil nuclear power reactor involving loss of life or serious damage. But as Sir Stanley Brown cogently argues in Chapter 3, it is better to be safe than sorry. The safeguards are stringent and they are likely to be relaxed with caution.

Beyond this, the fusion reactor, depending for its energy on the fusion of the light elements, remains a theoretical possibility, although its practical application has so far eluded the scientists. Its introduction as a power producer seems to lie far in the future. But it is there as an insurance against the depletion of other sources of energy, for the supply of the light elements is virtually inexhaustible.

If the impact of nuclear energy has been greatest in the military and strategic sphere, and most promising in the economic field as a source of power, it has been most immediate and most beneficial in medicine, industry, and agriculture.

The extraordinary value and variety of the applications in these fields is not often fully appreciated. Dr. Kronberger in Chapter 4 illustrates this theme.

As a source of power, nuclear energy is well adapted for the desalination of water and therefore provides an insurance against the depletion of water resources or an alternative, if the economics can be got right, to the construction of reservoirs and dams. In the form of radioactive isotopes, nuclear energy can also provide small power sources valuable, for example, in cardiac pace-makers, in spacecraft and, more mundanely, as a source of light for marine markers. But radioisotopes find their main use in medicine, industry, and agriculture; in medicine for research, diagnosis, and therapy; in industry for process control and the preservation of foodstuffs and materials; in agriculture and in mining for soil and product analysis.

As an indication of scale, the U.K. Radiochemical Centre at Amersham alone produces over 1400 different kinds of radioisotope and labelled compounds for medical use, and in the year 1969–70, had a turnover of nearly £4 million.

From the social standpoint, the consequences of the development of nuclear energy are more difficult to assess. Lord Ritchie-Calder discusses some of them in Chapter 6.

He illustrates the effect on the public mind of the way in which nuclear energy was developed and employed, and how its wartime origins have aroused primitive fears of the peaceful as well as the military applications. This has created an issue of confidence between scientist and layman, which, in spite of the excellent safety record of nuclear enterprises, has not yet been resolved. Indeed, the current concern with environmental problems has again stimulated public anxiety, especially in the United States. This anxiety is in general much exaggerated, but it has in some cases hampered and held up industrial projects which are otherwise desirable on economic and social grounds.

Finally, Lord Ritchie-Calder deals with the problem of the disposal or storage of nuclear waste. This also gives rise to public apprehension, but there are a number of acceptable solutions to it.

Nuclear technology has given a popular name to the age in which we live—the Nuclear Age. It has brought new benefits to society and with them new hazards; perhaps also it has increased the stresses and strains of contemporary life. It has certainly imposed new responsibilities on politicians and administrators for public security and safety, both at national and local level.

But in a broader sense, these stresses and strains, these new responsibilities are no more than one part of the technological revolution. It would be otiose to try and apportion merit or demerit to any one aspect of this revolution, which covers the whole spectrum of science and technology. Nuclear technology marks a distinct phase in this revolution, and if it has brought with it hazards as well as benefits, there is no cause, on this account, to be afraid of it. It is rather to be welcomed as another, rather dramatic, stage in the increasing mastery of man over his environment.

1

The science of the fission and fusion processes
By Robert Spence, C.B., F.R.S.

Radioactivity and atomic structure

It was established during the nineteenth century that matter consists of assemblies of atoms, and about ninety different atomic species or elements were identified, each chemically and physically distinct. Atoms are approximately a hundred-millionth (10^{-8}) of a centimetre in diameter; they respond to electric and magnetic fields, interact with light in various ways, and absorb heat by increased motion relative to one another. They combine together in groups or in extended arrangements according to fairly clearly defined rules to form the infinite variety of substances which constitutes the terrestrial world. In 1896 the French scientist Henri Becquerel noticed that when a crystal of a compound of uranium, then the heaviest known element, was left in a drawer over a photographic plate wrapped in its protective paper, an image was formed when the plate was developed. He thought that this *lumière noire*, as he called it, was connected with fluorescence. Shortly afterwards, however, Pierre and Marie Curie discovered a far more intense source of the radiation in uranium ore. They separated this element, which they called radium, from a large quantity of ore and showed the emission of radiation to be a new phenomenon known as radioactivity. It was found that emission from radium resulted in its transformation into another radioactive element and that there was a succession of such steps which ended at the element lead. Each radioactive species was shown to have its own characteristic rate of radioactive decay, usually represented by its half-life ($t_{\frac{1}{2}}$), the time taken for exactly half of the original material to become converted. Half-lives range from a fraction of a second to thousands of years or more and the number of disintegrations per second is expressed in curies (see Glossary). When it became established that radioactivity is independent of chemical form, temperature, or physical state, there could no longer be any doubt that this new phenomenon must be connected with some form of fundamental atomic substructure. The radiation coming from radium itself, called the alpha rays, was shown to consist of positively charged atoms of helium, the lightest element next to hydrogen, and Rutherford used these high-velocity particles to probe into other atoms. He found that in most instances they passed straight through the interior of the target atoms with no more than a slight deflection and only occasionally suffered a large deflection or recoil (Fig. 1.1). From the

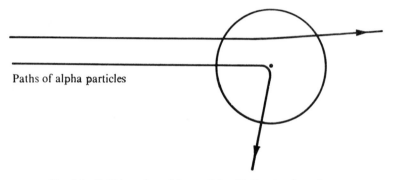

Paths of alpha particles

FIG. 1.1. Collision of an alpha particle with the atomic nucleus.

results of these experiments Rutherford calculated that practically the whole of the mass of the atom must be concentrated in a very small positively charged nucleus, of diameter about one ten-thousandth of that of the atom as a whole. He concluded that the region surrounding the nucleus is occupied by negatively charged electrons, sufficient in number to balance its positive charge. Electrons had been recognized previously as particles of relatively small mass (1/1800 of the mass of the hydrogen atom) carrying a fixed negative charge, and some years later Bohr showed that they are arranged in the atom in precisely defined layers or shells in a manner which reflects its properties in the Periodic System of the Elements. It is normally only the electrons in the outermost shell which are directly concerned in a chemical reaction. For example, when carbon burns in air, as in the combustion of coal, the outer electrons of the carbon atom and two oxygen atoms from the oxygen of the air become shared in a stable configuration and energy equivalent to $3 \cdot 28 \times 10^4$ joules per gramme (see Glossary) of carbon is released.

It became apparent that radioactive changes originating in the atomic nucleus are accompanied by the liberation of far more energy than is released in chemical change. The amount of energy given out by radium in a certain time was determined and from the result it was calculated that the complete radioactive decay of 1 gramme would produce 1.4×10^{10} joules, i.e. about half a million times as much as can be obtained from the combustion of 1 gramme of carbon. Three types of radiation from radioactive substances were distinguished, namely alpha rays, beta rays, and gamma rays. The alpha rays have already been described; beta rays consist of fast-moving electrons with a penetrating power roughly a hundred times greater than that of the alpha rays (e.g. 0·2 cm thickness of aluminium to stop beta rays as compared with 0·002 cm to stop alpha rays); gamma rays are a form of electromagnetic radiation of shorter wavelength than X-rays and can penetrate through several centimetres of lead. Rutherford continued to exploit the alpha rays in his investigations of the nucleus and in 1919 he showed that fast alpha particles from a radioactive substance could penetrate into and become absorbed by the nucleus of nitrogen. This was immediately followed by the expulsion of a proton and formation of an oxygen atom. The proton is the nucleus of the lightest atom, hydrogen, and carries a positive charge equal but opposite in sign, to the negative charge on the electron. This was the first time an atom was split in the laboratory; the reaction can be expressed in symbols as follows:

$$^{14}_{7}N + ^{4}_{2}He \rightarrow ^{17}_{8}O + ^{1}_{1}H.$$

(Here the superscript figure represents the mass number of the atom and the subscript figure its atomic number; for definitions see below.) Rutherford and his co-workers obtained similar results with other light elements but the alpha particles from radioactive sources were not sufficiently energetic to affect the nuclei of heavy atoms.

An important advance was made in 1932 when James Chadwick showed that various unexplained phenomena could be accounted for if one assumed the existence of another particle of about the same mass as the proton but electrically neutral, which he called the neutron. There were now three fundamental particles from which all matter was thought to be composed, namely the proton, the neutron, and the electron. Atomic nuclei could, in principle, be built up by adding protons and neutrons, beginning with the single proton of the lightest element, hydrogen, and ending with the heaviest element, uranium, containing 92 protons and 146 neutrons. Each chemical element is characterized by a specific number of protons in the nucleus, represented by Z, the atomic number, since chemical properties are determined by an equal number of electrons in the shells. The total number of protons, Z, and neutrons, N, in the nucleus is called the mass number, A. In the lighter elements, N is approximately equal to Z but as the atomic number increases, N becomes progressively greater than Z. There can be some variation in the number of neutrons associated with any particular number of protons and several atoms with slightly different mass numbers for the same Z value, display identical chemical properties. These are known as isotopes of the element in question (see Table 1.1). Most naturally occurring elements

TABLE 1.1

Nuclear composition of some important atoms

Symbol	No. of protons = atomic number Z	No. of neutrons N	Mass number A	Mass in atomic mass units
H	1	0	1	1·0081
D	1	1	2	2·0147
T	1	2	3	3·0170
He	2	2	4	4·0039
^{6}Li	3	3	6	6·0170
^{7}Li	3	4	7	7·0182
^{235}U	92	143	235	235·1170
^{238}U	92	146	238	238·1249

consist of several stable isotopes in fixed proportions. They can also have several radioactive isotopes, and Hevesy and Paneth were the first to show, as early as 1913, that radioisotopes can be used as 'tracers' for the element. By adding to ordinary lead a radioisotope of lead derived from the decay of radium, they were able to take advantage of the extraordinarily sensitive methods of detection and measurement of radioactivity to study the behaviour of lead in circumstances when other methods were either too insensitive or totally inapplicable.

The year 1932 was a memorable one for the Cavendish Laboratory in Cambridge. Besides the discovery of the neutron by Chadwick, Cockcroft and Walton succeeded, for the first time, in producing atomic transmutation entirely by artificial means. Protons were accelerated in a new kind of high-voltage machine to speeds sufficient to penetrate the nucleus, and in the first experiment with lithium as target, the atom which captured the proton split into two helium atoms:

$$^{7}_{3}\text{Li} + ^{1}_{1}\text{H} \rightarrow ^{4}_{2}\text{He} + ^{4}_{2}\text{He}.$$

Several hundred times as much energy is released in this process as is expended by the bombarding particle. Einstein had put forward the view, in 1903, that the mass of a body is a measure of its energy content and when the energy of a body is changed by an amount E, the mass of the body will change in the same sense by E/c^2 where c is the velocity of light, i.e.

$$E = Mc^2.$$

When the actual mass, as distinct from the mass number, of the reactants in a nuclear reaction is compared with the mass of the products, there is a difference which is found to be equivalent, in terms of the Einstein equation, to the energy released after allowing for the energy of the bombarding particle; for example, in the bombardment of lithium by protons,

	$^{7}_{3}\text{Li}$	$+ ^{1}_{1}\text{H}$	$\rightarrow 2\,^{4}_{2}\text{He}$	$+ Q$
	7·0182	1·0081	$2 \times 4\text{·}0039$	
Totals	8·0263		8·0078	Difference = 0·0185.

The figures are on the atomic mass scale, which is based on the convention that the mass of the isotope ^{12}C is exactly twelve mass units (Table 1.1). The loss in mass of 0·0185 units is, according to Einstein's mass-energy equation, equivalent to 17·2 MeV of energy (see Glossary), which is the same as the directly measured value; the energy of the bombarding proton was in this case negligible by comparison. It is possible, in a similar way, to calculate how much energy would be released if a given nucleus were to be formed from free protons and neutrons simply by obtaining the difference between the sum of the masses of the appropriate numbers of free protons and neutrons and the mass of the product nucleus. Dividing this total nuclear binding energy by the number of nucleons (i.e. protons and neutrons) present gives the average binding energy per nucleon. It turns out that this quantity is not the same for all nuclei but increases rapidly with increase of mass to a maximum value of 8·5 MeV at about mass number 40 and remains constant at this value until about mass number 120, beyond which it steadily decreases. In other words, nuclei with mass numbers lying between 40 and 120 are the most stable because they have the highest average binding energy per nucleon. This is clearly shown in Fig. 1.2. It implies that if light elements could be made to

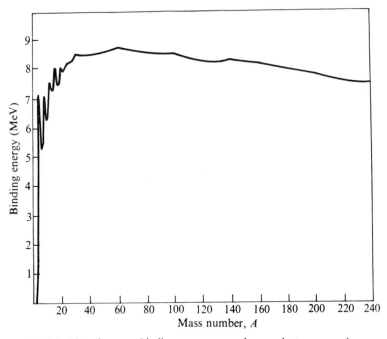

FIG. 1.2. Plot of average binding energy per nucleon against mass number.

combine to form heavier elements, a large amount of energy would be released. Unfortunately, bombardment by beams of energetic particles does not give a net gain in energy. For example, when the lithium atom was split by bombardment with fast protons, there was only one such event for every thousand million protons in the beam and the energy gain was negligible in relation to the total energy expended. The same considerations applied to bombardments in which light nuclei fused together and in the late 1930s there seemed to be no practicable route by which energy could be obtained from the fusion of light elements. It was recognized however, that such processes must occur in stars where the temperature is high enough for hydrogen nuclei (protons) to acquire sufficient energy to penetrate into and react with other light nuclei. The probability that a bombarding particle will react in some particular way with a target nucleus is defined in terms of a nuclear cross-section (symbol σ). This is the area which the target nucleus appears to present to the particle as determined from the collisions which are effective for the process under consideration. The unit of area used for nuclear cross-section measurements is called the 'barn'; it is equal to 10^{-24} cm², which approximates to the actual nuclear cross-sectional area.

Nuclear fission

In 1934 Enrico Fermi discovered that when an element is bombarded with low-velocity neutrons a radioisotope of the element is usually formed which decays by beta-emission to an isotope of the element next higher in atomic number. This led him to suppose that in the case of the heaviest element, uranium, with $Z = 92$, the process should result in the production of an isotope of a previously unknown element of $Z = 93$. He found that not one, but several radioactive species appeared to be formed, and concluded that they must be isotopes of 'transuranium' elements ($Z > 92$). Hahn, Meitner, and Strassmann in Berlin investigated the products of neutron irradiation of uranium in greater detail and advanced a more elaborate version of Fermi's theory. Then, in the summer of 1938, Lise Meitner, the physicist member of the team was unfortunately obliged to leave Germany. Hahn managed to keep in touch with her, however, and on 21 December 1938 he wrote stating that he and Strassmann had come to the conclusion, as chemists, that they were in fact dealing with radioisotopes of barium and lanthanum ($Z = 56$ and 57 respectively) rather than with isotopes of elements with Z in the neighbourhood of 92. Nuclear physicists had hitherto regarded such a drastic change as impossible or at least extremely unlikely, but Hahn was at that time one of the most distinguished radiochemists in the world and Lise Meitner felt that the observation must be correct. Her nephew, O. R. Frisch, had come to Sweden to spend Christmas with her and they quickly worked out an acceptable mechanism, in nuclear physical terms, for the process which they termed 'nuclear fission'. The announcement of Hahn and Strassmann's results and their interpretation by Meitner and Frisch caused something of an international sensation among nuclear physicists. The work was repeated and extended by scientists in many countries and by the summer of 1939 the main features of the process had become well established. It was shown that the two chief isotopes of natural uranium behave differently on exposure to neutrons of different energies. Slow 'thermal' neutrons are captured by the less abundant isotope ^{235}U (0·72 per cent of natural uranium) and the compound nucleus undergoes fission:

$$^{235}_{92}U + n \rightarrow {}^{236}_{92}U \rightarrow \text{fission.}$$

Neutrons of slightly higher energy are readily captured by the heavier, more abundant isotope ^{238}U (99·28 per cent of natural uranium) and in this case the compound nucleus decays by beta-emission with a half-life of 23 minutes to an isotope of a new transuranium element, subsequently called neptunium:

$$^{238}_{92}U + n \rightarrow {}^{239}_{92}U \rightarrow {}^{239}_{93}Np + e.$$

The energy released by the fission process can be calculated from the binding energy per nucleon as given in Fig. 1.2. This is about 8·5 MeV for atoms of medium Z and 7·6 MeV for uranium, so the total binding energy for the compound nucleus formed from ^{235}U by neutron capture is $236 \times 7 \cdot 6$ MeV and is $236 \times 8 \cdot 5$ MeV for the fission products. The difference, $236 \times 0 \cdot 9 = 212$ MeV, approximates to the total amount of energy produced as a result of fission, of which about 180 MeV is released almost instantaneously. Another important feature, which immediately distinguished nuclear fission from all other nuclear processes taking place in the laboratory, is that several neutrons are liberated. This gives rise to the possibility of a spontaneous and progressive spread of the process from the initial fission brought about by an extraneous neutron. In the case of ^{235}U the average number of neutrons released is 2·47. The excess of neutrons arises from the smaller neutron to proton ratio which prevails in the fission product elements as compared with uranium. It was realized that a chain of fission events would, if it could be sustained, be accompanied by a very large release of energy. For example, the complete fission of 1 kg of uranium would produce as much heat as the combustion of about 3000 tons of coal and if it were to take place all at once the effect would be equivalent to that produced by the explosion of 20 000 tons of TNT. After the outbreak of the Second World War, it was discovered that an isotope of element 94, subsequently known as plutonium, is produced when the neptunium isotope derived from ^{238}U decays by beta-emission with a half-life of 2·3 days:

$$^{238}_{92}U + n \rightarrow {}^{239}_{92}U \rightarrow e + {}^{239}_{93}Np \rightarrow e + {}^{239}_{94}Pu.$$

Plutonium-239 is a long-lived alpha-emitter (half-life 24 400 years), fissionable, like ^{235}U, by slow neutrons.

Chain-reacting systems

There was at first some doubt whether the fission chain could propagate in natural uranium since the fissionable isotope ^{235}U is present only to the extent of 0·72 per cent. An essential condition for this to happen is that, on the average, at least one neutron formed in the fission process will bring about another fission. The probability of neutron capture by ^{235}U can be greatly increased by reducing the speed of the fission neutrons to thermal velocities as soon as possible. This can be done by surrounding the uranium with carbon (as graphite) or deuterium (an isotope of hydrogen, mass number 2, in the form of heavy water); the neutrons quickly lose energy by collision with the carbon or deuterium atoms. It was found that the most efficient arrangement is a lattice structure consisting of bars of uranium surrounded by the moderator of graphite or heavy water. If the system is below a certain size, so many neutrons escape from it that there are too few left to propagate the chain. However, the proportion of external losses is reduced as the size increases and, provided internal losses are not too great, a chain reaction will develop when the system exceeds a certain critical size. Neutrons are lost internally, so far as chain propagation is concerned, through capture by ^{238}U to give neptunium and ultimately plutonium, through capture by atoms other than uranium such as the moderator and impurities, and at a later stage, through capture by certain fission products. Boron has a particularly large capture cross-section for slow neutrons and must be rigorously removed. Neutron capture by ^{238}U can be optimized in relation to ^{235}U fission by adjusting the thickness of the bars and the ratio of moderator to uranium. As the speed of propagation of the fission chain reaction is quite rapid, it might be expected that it would be difficult to control. Fortunately the release of a small fraction (0·75 per cent from ^{235}U) of the neutrons produced in fission is subject to delay; they arise from some of the short-lived fission products which decay with half-lives ranging from a fraction of a second to about a minute. This allows ample time for the neutron population to be controlled by the mechanical insertion or removal of a neutron absorber. In a practical system, neutrons must be sacrificed to absorption by structural materials and coolant and the critical size becomes large; the core of the graphite-moderated research reactor BEPO at Harwell, for example, is a cylinder 6·1 metres in diameter and 6·1 metres long.

It became clear at an early stage that the possibility of obtaining a more intense source of nuclear energy, sufficient to cause an explosion, depended on obtaining the fissile isotopes ^{235}U or ^{239}Pu in concentrated form. Both of these tasks were accomplished in the United States during the war. It was known that when two gases of different density are allowed to diffuse through a porous membrane, the lighter gas passes through more quickly because the rate of diffusion of a gas is inversely proportional to the square root of its molecular weight. The only convenient gaseous form of uranium is its hexafluoride and the difference between the square roots of the molecular weights of ^{235}UF$_6$ and ^{238}UF$_6$ is extremely small. However, by using a very large number of membranes and by moving the light and heavy fractions obtained by partial diffusion at each membrane in opposite directions in the plant, pure ^{235}U was eventually obtained. In the case of plutonium, large nuclear reactors or 'piles' were built using the system of natural uranium rods with graphite moderator, and plutonium was allowed to accumulate in the rods. These were removed after a time, dissolved in nitric acid, and the plutonium recovered in pure form from the uranium and highly radioactive fission products by a chemical process.

Once pure plutonium or highly enriched uranium is available, it is no longer necessary to rely on slow neutrons for propagation of the chain since the probability of fission induced by fast neutrons is then adequate. In these circumstances the moderator can be dispensed with and since there are few impurities and no ^{238}U present to compete for neutrons, the critical size becomes quite small (in the region of 15–20 cm diameter). If a supercritical mass of ^{235}U or ^{239}Pu can be assembled, the fast neutron chain reaction will then take place extremely rapidly in an uncontrolled fashion, and provided the system can be held together long enough for a substantial fraction of the material to react, there will be an immediate production of a very large amount of energy with explosive effect. It is also possible to use highly enriched uranium or plutonium in a reactor arrangement in which the fast neutron chain reaction can be controlled. This system, known as a fast reactor, contains no moderater other than structural materials and coolant, and is relatively small in size compared with thermal neutron systems (core diameter less than 1 m).

Considerable benefits can also be derived for thermal neutron systems from slight enrichment of ^{235}U in natural uranium, for example, up to 2 per cent. This allows a wider range of structural materials to be used and the substitution of ordinary water for heavy water as moderater.

Theory of fission

Meitner and Frisch first described the fission process in terms of the liquid drop model of the nucleus put forward by Bohr in 1937. Their ideas still retain a measure of validity though they have been elaborated in various ways in the light of later knowledge. Individual nucleons, whether protons or neutrons, are held together in the nucleus by powerful attractive forces which fall off extremely rapidly with distance. There is at the same time, a repulsive force between the protons due to their positive charge, which follows the ordinary electrostatic law, and is operative at distances at which the attractive force has become negligible. The effect is that a proton approaching the nucleus will experience an increasing repulsive force until it reaches a point when the short-range attractive force rather suddenly becomes dominant. Nuclei such as ^{235}U, which contain a large number of nucleons, may therefore be considered as being analogous to a liquid drop in which the attractive forces between the molecules of the liquid tend to draw it together into a form which leaves the minimum of unsatisfied forces at the surface, i.e. a sphere. When the nucleus captures a neutron, energy equal to the binding energy is liberated and the nucleus becomes temporarily excited, undergoing various distortions. Normally the surplus energy is emitted as gamma radiation and the shape returns to the stable spherical form. In the case of atoms of high mass number such as ^{235}U, containing a large number of protons, a point may be reached where the surface force tending to restore the spherical shape is overcome by the repulsion between protons in the opposite portions of the distorted nuclear drop. It then splits into two smaller nuclei which fly apart at great speed while some of the surplus neutrons are freed at the same time (Fig. 1.3). Nuclei like ^{235}U and ^{239}Pu generate sufficient internal

FIG. 1.3. Stages of nuclear fission according to the liquid drop model.

excitation energy, even after absorbing a neutron of very low energy, to bring about the crucial distortion, whereas ^{238}U requires a contribution from the kinetic energy of the bombarding neutron of at least 1 MeV. This difference is partly attributable to the greater binding energy obtaining when a compound nucleus is formed with an even number of neutrons since it has been found that there is a tendency for nucleons to form stable pairs.

It might be expected, on the basis of the simple liquid drop model, that the compound nucleus would split into two equal fragments, or that the fission product mass numbers would at least be symmetrically distributed, statistically, about the equal mass number mode. This is in fact what tends to happen when high-energy bombarding particles are used, but with slow neutrons the distribution of fission product mass numbers is asymmetrical, the most probable values being $A = 95$ and $A = 139$ rather than $A = 234/2 = 117$. The fluctuations which appear in the plot of the average binding energy against A (Fig. 1.2) give some indication, however, that the nucleus cannot be considered as a structureless drop and there is now a great deal of other information which points to the existence of a shell structure. This has been taken into account in various refinements of the theory of nuclear fission but the problem is a complex one and is still not entirely solved.

The products of nuclear fission

After fission has taken place and some neutrons have been liberated, the two primary fission fragments still retain more neutrons than required by the stable isotopes corresponding to their particular mass numbers. In a few cases, these are partly released as the delayed neutrons which have already been referred to, but otherwise, beta particles are emitted in succession until a stable isotope is reached. This leads to the formation of a chain of fission products of which the following is a noteworthy example:

$$^{140}_{54}Xe \xrightarrow{16\,s} e + {}^{140}_{55}Cs \xrightarrow{66\,s} e + {}^{140}_{56}Ba \xrightarrow{12\cdot8\,d}$$
$$e + {}^{140}_{57}La \xrightarrow{40\,h} e + {}^{140}_{58}Ce \text{ (stable)}.$$

On the average, however, there are only three stages of beta-emission per chain. More than 60 such chains have been identified containing over 200 different radioisotopes of 35 elements with mass numbers between 72 and 160. The great majority fall into two groups, a light group with $A = 85$–104 and a heavy group with $A = 130$–149, and half-lives mostly range from a fraction of a second to about thirty years. Energy is released by the fission products in the form of beta particles, neutral particles of extremely small mass even compared with the electron, called neutrinos, and gamma radiation. The neutrino is emitted simultaneously with the beta particle, carrying with it a share of the energy. It penetrates through matter extremely easily and so the energy which it carries is very widely dispersed. The energy of fission is therefore divided into two parts:

(a) released immediately as
    (i) kinetic energy of the fission fragments 165 MeV,
    (ii) energy of prompt neutrons 4·9 MeV,
    (iii) energy of instantaneous gamma rays 7·8 MeV,

and (b) released from the fission products over a period determined by their half-lives,

    (i) energy of beta particles 9 MeV,
    (ii) energy of gamma radiation 7·2 MeV,
    (iii) energy of neutrinos 16 MeV.

As the early members of the fission product chains tend to have short half-lives, the total beta and gamma emission diminishes very rapidly at first and then more and more slowly with time. After a 5–10-fold reduction in the first 60 days, it continues to decrease for over 100 years. The beta activity will be less than a hundred-millionth of the initial value after 500 years but it will change extremely slowly beyond that owing to the presence of a few fission products with very long half-lives, such as ^{99}Tc with $t_{\frac{1}{2}} = 2 \cdot 12 \times 10^5$ years. In a practical system, the ^{239}Pu formed by neutron capture from ^{238}U is separated from the fission products by a chemical process but isotopes of other transuranic elements formed by successive neutron capture usually remain with the fission products. Some of these have half-lives of a few hundred years, such as $^{241}_{95}$Am ($t_{\frac{1}{2}} = 458$ years) and thus add to the problem of long-term radioactivity of nuclear wastes.

Just as plutonium can be obtained from ^{238}U by neutron capture followed by beta decay, it is also possible to obtain a fissile isotope of uranium by neutron irradiation of the element thorium:

$$^{232}_{90}\text{Th} + \text{n} \longrightarrow {}^{233}_{90}\text{Th} \xrightarrow{23 \cdot 5\,\text{m}} \text{e} + {}^{233}_{91}\text{Pa} \xrightarrow{27 \cdot 4\,\text{d}} \text{e} + {}^{233}_{92}\text{U}.$$

The intermediate isotope of protactinium has a half-life about ten times longer than that of neptunium-239, which means that the build-up of ^{233}U occurs rather slowly to begin with. Thus there are a number of alternative routes to the release of useful energy from fission, based on the three fissile isotopes ^{233}U, ^{235}U, and ^{239}Pu.

Nuclear fusion

When an uncontrolled nuclear chain reaction occurs in pure fissile material, as in a bomb, temperatures about the same as those which prevail in some stars are attained and this opened the possibility of obtaining further energy from fusion of light elements. In order for fusion to take place, it is necessary for the nuclei to approach one another with sufficient speed to overcome the repulsion arising from their positive charges. Because heavier nuclei have higher positive charges they exert a greater repulsive force and consequently fusion should take place more readily with the isotopes of hydrogen than with nuclei of higher atomic number. Although fusion reactions involving protons proved to be too slow at the attainable temperatures, reactions of the other hydrogen isotopes, deuterium (D) and tritium (T), are fast enough to be of practical interest. These are as follows:

$$^2_1D + ^2_1D \rightarrow ^3_2He + ^1_0n + 3{\cdot}27 \text{ MeV},$$
$$^2_1D + ^2_1D \rightarrow ^3_1T + ^1_1H + 4{\cdot}03 \text{ MeV},$$
$$^3_1T + ^2_1D \rightarrow ^4_2He + ^1_0n + 17{\cdot}6 \text{ MeV}.$$

The D–T reaction is about a hundred times faster than the D–D reactions and yields more energy. Tritium is a radioactive gas which emits beta particles with a half-life of 12·26 years and is made by bombardment of lithium with slow neutrons in a nuclear reactor. Lithium consists of two isotopes (7·5 per cent ^6_3Li and 92·5 per cent ^7_3Li) only one of which yields tritium:

$$^6_3Li + ^1_0n \rightarrow ^4_2He + ^3_1T + 4{\cdot}6 \text{ MeV}.$$

In order to bring about the fusion of deuterium and tritium by means of the heat generated from fission it is desirable that they should be present in as concentrated form as possible. Both are gases at room temperature but the deuteride (or tritide) of lithium, LiD, is a convenient solid compound, and by using the separated lithium-6 isotope tritium can be generated *in situ* during the reaction. In such a system, the amount of energy liberated per gramme is even greater than is obtained from fission.

The possibility of obtaining a controlled release of fusion energy has attracted a great deal of attention in recent years. At the temperature of fusion, the outer electrons of the atoms become separated from the nuclei and the system is then called a 'plasma'. Plasmas can be created in the laboratory at lower temperatures and since they are electrical conductors and responsive to magnetic fields, it is hoped to compress magnetically or otherwise raise the temperature of a deuterium–tritium plasma to such an extent, and to contain it for sufficient time, as will allow fusion to occur. Although considerable progress has been made in this direction, success has not yet been achieved.

Effects of nuclear radiation on materials and on living organisms

All processes for the large-scale release of nuclear energy generate various forms of radiation which give rise to serious problems in practical applications. Nuclear particles and gamma rays usually carry a much larger amount of energy than is required to break an ordinary chemical bond and when they pass through matter they lose their energy progressively by displacement of electrons (i.e. by ionization), by electronic excitation of atoms and molecules, and by displacement of atoms from their position in a chemical compound or crystal lattice. The amount of radiation absorbed is expressed in Röntgen units (symbol R; see Glossary). Simple substances such as very pure water, carbon dioxide, and uranium dioxide are remarkably stable to radiation because of processes which operate automatically to restore the original state of affairs; in the case of more complex substances, the damage tends to be irreversible. Neutrons suffer elastic collisions with atoms which may then be expelled from their stable position in a compound or crystal lattice by recoil. Hydrogen atoms, for example, are expelled from hydrogenous molecules as protons, and carbon atoms in graphite are driven out of their crystal layer plane into an interlamellar position, causing the crystal to grow in one direction and shrink in the other. Neutrons give rise to the further complication in that when captured a radioisotope is normally produced which may be long-lived. This means that structural materials, such as steel, when exposed to radiation in a nuclear reactor remain radioactive for a long period after removal. On the other hand, neutron capture processes are the source of valuable supplies of radioisotopes for use in research, medicine, and industry.

Living organisms also suffer radiation damage. Heavy doses, especially if delivered all at once, cause the individual cells either to die or to suffer a loss in efficiency, and this leads to malfunctioning of the organism as a whole and possibly to death. In human beings a dose of 500 R to the whole body results in death in over 50 per cent of cases. Lower doses may give rise to leukaemia, to some forms of cancer, and to various other disorders, and in addition to these somatic effects there may be damage to the germ cells. Genes in the chromosomes of germ cells, though remarkably stable, suffer occasional changes known as mutations which may lead to the appearance of detrimental genetic traits, such as haemophilia, in the progeny. These changes occur naturally and it is thought that the radiation from cosmic rays and from natural radioisotopes to which all living organisms are exposed, amounting to about 3 R for human beings, is responsible for a proportion of such mutations. Information concerning human beings is very uncertain since it is based almost entirely on experiments with plants and animals, but it has been estimated that an average dose of 30–80 R over the whole population would be required to double the natural mutation rate. If the hazards of nuclear war are excluded, the probability that more than a small fraction of the population could receive a dose of this magnitude is extremely small. On the other hand, as any increase in the average dose to the population, however small, must be assumed to lead to some increase in the rate of mutation, it is generally accepted that this must be avoided or at least minimized. The high sensitivity and accuracy of nuclear measuring equipment permit an exceptional degree of control to be exercised and there are now internationally agreed maximum permissible doses, but constant vigilance will always be necessary.

Glossary

curie   the unit of radioactivity. $1\ C = 3 \cdot 7 \times 10^{10}$ disintegrations per second.

joule   the unit of energy. The temperature of 1 cubic centimetre of water is raised $0 \cdot 24\,°C$ by 1 joule of thermal energy.

MeV   million electronvolts. 1 eV is the energy acquired by an electron or other particle of the same charge moving though a potential gradient of 1 volt. $1\ MeV = 1 \cdot 602 \times 10^{-13}$ joule. If each atom of an element emits 1 MeV, the atomic weight in grammes would yield an amount of energy equal to $9 \cdot 64 \times 10^{10}$ joules.

röntgen   the unit of radiation dose. 1 R is the quantity of X or gamma radiation which will produce in 0·00129 gramme of air (1 cubic centimetre of dry air at 0 °C and 760 mm pressure), ions carrying 1 electrostatic unit of electricity of either sign. The energy gained by the absorption of 1 R per gramme is equal to about 10^{-5} joule.

electron   symbol e, charge −1, rest mass in atomic mass units 0·00055.

proton   symbol p, charge +1, rest mass in atomic mass units 1·00728.

neutron   symbol n, charge 0, rest mass in atomic mass units 1·00867.

neutrino   symbol ν, charge 0, rest mass in atomic mass units zero or very small.

2

The technological development of the uranium and 'hydrogen' bombs

by Air Chief Marshal Sir Denis Barnett, G.C.B., C.B.E., D.F.C.

Initial summary

The account in Chapter 1 of the science of the fission and fusion processes shows that these processes can be made to yield outputs of energy so great that the mind has to make a deliberate effort of re-attunement in order to begin to comprehend their significance. It is also evident that, while this scientific knowledge was unfolding to reveal possible uses for such huge potential outputs of energy, one foreseeable use would be for war: or, as has less foreseeably happened, for the prevention of war.

In the event, because Europe was already at war when the advance of scientific knowledge had disclosed these possibilities, the use toward which endeavour was directed was the warlike one—the production of a nuclear fission war-head (an 'atomic bomb') to be carried by aircraft. The first of these (a test device) was exploded from a tower at Alamogordo in New Mexico, U.S.A. in July 1945. The second and third (the only two to be used in war) were delivered upon their targets at Hiroshima and Nagasaki in early August 1945. Some seven years later (November 1952) the first fusion device ('hydrogen bomb') was tested by the United States. The U.S.S.R. and the U.K. exploded their first fission devices in September 1949 and October 1952, followed by 'hydrogen bombs' in August 1953 (the U.S.S.R.) and May 1957 (the U.K.). Later, in February 1960 and October 1964 respectively, France and China fired the first of their tests in the series which each has since continued.

The progression by which evolving international technology has reached and passed these milestones is the subject of this chapter.

The starting-point

We begin at the point when existing laboratory techniques (using minute quantities of fissile materials) first began to be developed into a war project on a national, and later an international, scale. To define just when that transformation began is to make a subjective choice of starting-point, but a practical one is to take the point of time at which the question 'can it be done ?' was, under the spur of war, first underpinned (or overtopped) by a decision to go ahead and do it. Four dates in particular bore upon that decision and upon its fulfilment: February 1940, April 1940, November 1941, and December 1941.

In February 1940, a short scientific note, which has come to be known as the Frisch–Peierls Memorandum, set forth a series of predictions which were to change decisively the whole direction and impetus of scientific thought on this topic and were to set in train the first endeavour in Britain toward a war project to develop a fission bomb. Those predictions led directly and quickly to the setting-up in April 1940 of Professor Sir George Thomson's Committee, soon to become known on both sides of the Atlantic as the Maud Committee. For 18 months the Maud Committee steered Britain's effort and collaboration with the U.S.A. and with Canada and France until, in November 1941, it led to the nomination of the British project team known by the code name *Tube Alloys*. In December 1941 came Pearl Harbour. Thereafter, under the war-time code title *Manhattan Project*, the vast resources of the U.S.A. fastened upon the task with an outcome, 43 months later, about which the world first became aware just before it learned also of the ending of the war.

So much for the broad sequence of events. Let us now briefly recall what it was in the Frisch–Peierls Memorandum that was so compulsive in its influence upon thought in February 1940.

Speculation about the possibility of a nuclear fission bomb had up to that time been in terms of natural uranium as the fissile constituent. Most of the conclusions which had emerged from that line of thought had been negative. Specifically, if in oversimplified terms, two conclusions seemed to rule it out. A practical one was that the critical mass required for a chain reaction in natural uranium would be so large (100 tons) as to be self-evidently inapplicable to any bomb which an aircraft could carry. Secondly, a chain reaction would not proceed fast enough in natural uranium to be effective as a bomb device. The reaction, if it could be started effectively, would be overtaken by its own disruptive effect upon the critical mass, whereupon the reaction would stop, having yielded no more than a fraction of its energy broadly comparable with that from conventional explosives.

It was Chadwick, Frisch, and Peierls who perceived how dramatically different might be the behaviour if the preponderant isotope ^{238}U were separated out from natural uranium and if the resultant highly enriched ^{235}U were employed instead. Although the separation of uranium isotopes was known to be formidably difficult, even on a minute scale in the laboratory, the Memorandum predicted that a separation plant could be developed to the scale needed for quantity production if enough effort were harnessed to the task. Thus it was tenable to consider what might be the characteristics of a bomb with a fissile charge consisting of almost pure ^{235}U. That proposition, as the authors pointed out, had not yet been seriously considered, nor had the behaviour of the isotope ^{235}U under bombardment by fast neutrons yet been established experimentally. But the authors of the Memorandum predicted that almost every collision of a neutron with a ^{235}U nucleus would produce a fission, and that neutrons of any energy would be effective.

From those predictions there flowed two more of prominent importance. First, that the neutron multiplication would be such that the critical mass of the fissile material in a bomb could be of manageable size, measurable in kilogrammes rather than in tons. Secondly, because fast (as well as slow) neutrons would be effective, the chain reaction would proceed with extreme rapidity—so fast that a substantial fraction of the total energy would be liberated before the material had had time to blow itself apart into sub-critical masses and so stop the reaction.

Further predictions in this remarkable document covered such aspects as the likely size for a practical bomb, a method of initiating its explosion, and an estimate of its behaviour and yield.

But, without completing the catalogue of these penetrating (and broadly correct) conclusions which set the initial course ahead, it might be as well to turn aside at this stage to think about the dimensional significance of some of the expressions already used in this account in order to grasp more readily the dimensions of others which will follow. Without a continuing effort of mind it is not easy to comprehend the reality of the scientific and technological endeavour which the bomb project was to demand.

26

To take one random example: when the Memorandum asserted that a chain reaction in ^{235}U would proceed very fast, it was speaking of the sort of time interval which is defined not in seconds or in thousandths but in millionths of a second. Moreover, it some contexts, it can become necessary to subdivide still further, for important sub-events can occur within a main event at intervals which may have to be defined in nanoseconds. (A nanosecond is one-thousandth of one-millionth (1×10^{-9}) of a second.) To establish what has happened (and sometimes even more importantly what has not happened) during a short interval like that demands a combination of very advanced techniques of prediction and of verification. To begin to understand why a sub-event has happened or not happened demands yet another, still more diversified and advanced; and so on.

Throughout this novel project the way ahead had to go far beyond conventionally understood boundaries towards new extremes of environment within which the behaviour of new materials had to be predicted and understood. For example, in terms of temperature the range to which the mind must address itself is from near absolute zero $(-273\,°C)$ to something like $10^{10}\,°C$; in terms of pressure, to something like 10^{13} atmospheres; in terms of purity, to the regulation of contaminant down to about 1 part in 10^5; and so on.

A further example, a more sternly practical one, is provided by the production of ^{235}U in effective quantity. The assertion in 1940 that it would be possible to devise industrial-sized plant capable of doing this was ultimately proved correct. By mid-1945, after 5 years of endeavour on both sides of the Atlantic, it had been done: but only just. The margins both of timing and of certainty (or uncertainty) had been close-run indeed. Meanwhile, during the intervening five years, the expenditure of effort and money had reached very high levels. Because of the extreme technical difficulties which were encountered as the work proceeded, the Americans had deemed it essential, in order to insure against failure, to pursue simultaneously no fewer than three separate technological routes towards the production of ^{235}U. And, because of the margins of uncertainty which beset even that proliferation of effort, they also pursued a different route towards an entirely different fissile material (plutonium) in order to have this alternative for use in a fission bomb if the difficulties of producing ^{235}U in effective quantity should prove too great. In the event they succeeded in producing both—but, even under the urgent impetus of war, only just.

The prodigious scale of those exertions provides an illustration of what each stage in the development of the bomb was to demand. It is particularly useful to reflect upon the vast output of development effort that had to be devoted to one step alone, namely the production of the fissile material.

The early steps

Let us now pick up the thread again at the point (April 1940) when the Maud Committee went to work on the conclusions of the Frisch–Peierls Memorandum. We have already seen that the content of that almost uncannily prescient document was unexampled in its grasp of the matter at that time. It was also a document of exquisite conciseness: no more than a few pages of typescript. Fortunate indeed was the Maud Committee to have it as the initial feed-stock for its deliberations.

From the work of the Maud Committee there emerged very considerable findings, but these must, for our purpose, be abruptly compressed. The cardinal one was the clear, affirmative answer which the Committee gave to the question 'can it be done?' To the two major corollary questions 'if so, how?' and 'what must first be done?' the Committee also gave its answers. Its contemplated bomb was to contain some 10 kilogrammes of ^{235}U divided into sub-critical component fractions which would be held apart within the bomb until detonation was required. To initiate the chain reaction, a gun-type mechanism would fire the sub-critical portions together to form a critical mass. It was recognized that the speed of assembly into critical mass would be of crucial importance. With the knowledge available at the time, it seemed that the achievable rate of closure of the sub-critical masses would be about 6000 feet per second and it was estimated that the explosion resulting from assembly into critical mass at that rate might produce a yield of energy equivalent to that from 1·8 kilotons of conventional explosive.

To the second question 'what must first be done?' the Committee's answer correctly diagnosed the prominence of the need to produce ^{235}U in workable quantity and the prominence also of the severe difficulties which would be encountered in doing this. We have already seen how right that diagnosis proved to be and how right was the Committee to put its finger upon isotope separation as being something on which work must start at once. This was not to imply that everything else could wait. There was a thicket of requirements attached to every single one of the legion of scientific and technical topics implicit in the design of the bomb. Prominent across the whole spectrum stretched the particular need to develop a quickening programme of collaboration with counterpart work in the U.S.A. For underlying everything was the need for an ever-widening and more and more penetrating scientific understanding.

Instead of pausing here to follow up seriatim the detail of these tasks, it is more useful now to move ahead in time and to move also across the Atlantic (whither the project teams had been shuttling to and fro) in order to take stock of the gigantic programme which had been set in train after Pearl Harbour and was being tackled by the Manhattan Project.

The middle steps

A convenient date to pick for such a stocktaking is the turn of the year 1944–5. That date may seem over-far ahead as a mid-point, because the successful outcome then lay ahead only a matter of months in time, but it was much less near in terms of what still had to be done. It is true that prodigies of achievement had filled the three preceding years and huge agglomerations of plant had been constructed. But significant parts of all this had not long been commissioned into effective production. The really refractory obstacles had remained refractory and not for years had they begun to yield. It was not, for example, until late in 1944 that there was developed and produced in quantity (by new techniques of powder metallurgy) an effective barrier membrane for use in the vast gaseous diffusion plant which had been built for the separation of ^{235}U. That quest, it will be recalled, had been pursued with fluctuating prospects from the very outset.

So it had been also, though in differing degrees, with the other routes to production of ^{235}U. Of these the Manhattan Project had originally reconnoitred four. One of them, the centrifuge, had been set aside because not enough basic work had by then been done on it. Indeed it is only very recently, after nearly 30 years, that it has re-emerged into the public view. Another route, by the process of thermal diffusion, was similarly set aside, but more temporarily. It was taken up again later (about mid-1944) and a plant was constructed which began to produce at low enrichment at the turn of the year. Its product was put to use as a beneficially semi-enriched feed-stock for further enrichment via the other two routes. Of these one was the gaseous diffusion process already mentioned. The other was the process of electromagnetic separation. These two were the chosen mainsprings of ^{235}U production and to them had been devoted effort upon the grandest scale. Yet success had remained capriciously elusive, particularly via the gaseous diffusion route. For something like a year the electromagnetic separation route was the only one operating effectively and, indeed, until some 18 months after the end of the war, that route was still the only one capable of final enrichment to ^{235}U of weapon grade.

And so, at the date of our selected viewpoint at the turn of 1944–5, the manufacture of ^{235}U in quantity was still poised at the threshold of success.

We have seen that it was because of these uncertainties confronting ^{235}U production that the decision had been taken to produce quite another fissile material (plutonium) for use, as necessity might dictate, in the bomb. For plutonium was a material wherein the contaminants that would have to be removed to reach an acceptable purity would be the chemical contaminants, not the isotopic ones. And so, though the overall method of production would be a process of formidably massive scale, demanding the construction of a complex of nuclear reactors to provide the necessary irradiation, the culminating separation could proceed by the more straightforward and fairly well-established techniques of industrial chemistry.

At the time the decision was made it was necessary to decide also what type of reactor to develop for the purpose and, in particular, whether the moderator would have to be heavy water, as the earlier work on such reactors had suggested, or whether graphite could be used instead. The pursuit of either course could be seen to require much development, quite apart from the construction of the reactors themselves. For the former it would be necessary to build new plant for producing heavy water in the necessary quantities. For the latter it would be necessary to undertake extensive research in order to develop effective means of regulating the behaviour of graphite under irradiation.

In the event the decision went in favour of graphite as the moderator, but the uncertainty had persisted beyond the time when it had become necessary to go ahead with building heavy-water plant in case it might prove essential. So here again, for part of the time at least, multiple routes had to be followed. Aggregating these with their tortuous counterparts in the ^{235}U production line one begins to discern some measure of the daunting scope of development upon which the Manhattan Project had engaged.

The vast amount of activity which had been directed to what might be called the fissile heart of the matter tends, by its sheer size, to dwarf the complementary developments which were needed to make that fissile heart work. It would be impossible to list in any useful perspective (without inordinate tedium) the array of other scientific and technical topics upon which the project had, in one way or another, to draw and to develop far beyond earlier practice or even understanding. But, taking just one of those peripheral topics by way of example, let us consider in outline what had been required to develop the means of assembly of sub-critical masses of fissile material into critical mass to initiate the explosive chain reaction. At the outset of the project it had been propounded that this might be done by a gun-type mechanism. The four ensuing years of development produced such a mechanism. But that type of assembly system, though it was a practicable one for use in a ^{235}U bomb, was not applicable to a plutonium bomb. For the latter a far higher speed of assembly was required than could be achieved by any type of gun. What was needed was a mechanism of implosion whereby the energy from conventional high explosive could be directed inward not outward. From a peripheral structure of high explosive the energy had to be directed inward upon an enclosed mass of plutonium of sub-critical density. The performance required was that of squeezing the sub-critical mass into criticality with near-instantaneous effect. If assembly were any slower than that, a portion of the fissile material would pre-detonate before full criticality and the bomb would fizzle instead of detonating effectively. Moreover it was not just a matter of rapidity. The implosion must also proceed symmetrically. The rigour of those demands was such that they carried this important sub-project (just as other demands had carried other sub-projects) right out into previously unexplored fields of technique and of underlying physics. Indeed, it was not until late in 1944 that it could be asserted with confidence that this crucial element of development would succeed.

Because of all this, the choice of 1944–5 as the point at which to take stock has been probably as serviceable as any other. Surveying the maturing assets which had emerged from the grind of the intervening years, one could by then at last perceive that the timing which lay ahead would almost certainly be governed only by the rate of accumulation from the separation plants of enough of the two fissile materials for the fabrication of the first bombs. And so it was to prove.

The next middle steps

Successful production of the first fission bombs was not, however, a final step. It was not even the end of the middle. A new immensity lay next ahead.

In Chapter 1 we saw that the potential output of energy which can be obtained from a fusion reaction is of a magnitude quite different from that obtainable from fission: different not just in the comparative sense of weight for weight of reacting material, but different also in the total potential output. This is so because practical (as well as theoretical) considerations of criticality limit the amount of fissile material which can be held poised for implosive assembly within a fission bomb. There is theoretically no limit to the amount of fusion reagent which can be held ready for fusion, for that reaction cannot begin until the material is effectively heated to a quite abnormal temperature. The temperature needed lies in the million-degree range which is characteristic of some stars and which, on our planet, requires an explosive fission reaction to generate it. Until July 1945 that means did not exist. But from then onward it became possible to develop (not just to contemplate) devices yielding outputs of energy measured in megatons of TNT equivalent instead of in kilotons.

Because the mere possession of such devices, however powerful, is a limited asset unless it also includes possession of the means for their delivery, it becomes important from now on not to overlook the vast additional field of technological development called into being by the need for a delivery system. In what has gone before it has been convenient to ignore it because the devices were then delivered by conventional aircraft that were already in existence. The state of the art of delivery now could hardly be more different. Yet, despite the interest of that progress, it would not be convenient to turn aside to discuss it: the war-head topic is more than big enough by itself. But we need to retain a sense of the doubled extent of this gamut of development as an entity. Perhaps one way of doing this is to reflect that, as just one among the many constituent parts of the whole, the entire panorama of space technology, including man's transit to the moon, can be labelled as an excursion from the main stream of technology generated by the bomb project; so also can nuclear power stations, radioisotopes, and much else in modern everyday life.

We now revert to the fusion reaction which constitutes the kernel of the so-called 'hydrogen' bomb. The statement (two paragraphs back) that that reaction cannot start unless the material is effectively heated was deliberately made in these particular words because it is not just a matter of applying heat. An essence of the process is that it must be effectively contained during the build-up of the extremes of temperature and pressure generated by the fission trigger. Moreover, we have seen in Chapter 1 that one of the characteristics of some fusion reactions is that they can produce neutrons of very high energy capable of causing fission in ^{238}U. Thus we must picture a further development from a fission–fusion device to a fission–fusion–fission one. The total yield of energy is then incomparably more economical since the final (fissile) reagent can consist of the preponderant isotope of natural uranium without the necessity for tedious and costly separation.

It will be apparent that a prominent feature of either of those developments is that the device, at initiation, creates within itself a local environment of uttermost violence. It will be equally apparent that, in order to secure an orderly functioning of components in circumstances so novel and so harsh, the design criteria must themselves be of a stringency so unexampled as to demand the development of yet another range of new technologies to match them. Indeed it is by reflecting upon these extremities of circumstance as a whole that one can best comprehend the size of this phase of the project. The account of the earlier phases provides a framework for this, but the differences mostly far outweigh the analogies and the techniques more often have had to be new ones than refinements of the old.

The whole, complicated yet further by the great array of paraphernalia developed for conducting tests underground in compliance with treaty obligations during this phase, has had to embrace a breadth of technological progress almost undreamt of (because uncalled for) before. That progress, despite its considerable achievements since 1940, is still in mid-stream.

Between World War I and World War II there intervened 21 years of uneasy peace. Of this commodity our quota since then is already more than that despite comparable threat and danger. Each one of us can hold his own opinion about why that has been so.

3

The nuclear power programme

By Sir Stanley Brown, C.B.E.

It has been said that the intensity of power usage is a measure of the degree of advance of civilization. Certainly early civilizations were based on muscle power—either slaves or domestic animals, with their inherent limitations—and broadly speaking that held until the invention and improvement of mechanical power accompanied the Industrial Revolution in the late eighteenth and nineteenth centuries. Since then, with the complexity of our civilization and improvement in the general standard of living, the demand for power has grown steadily until at present the total demand for power in all its forms is increasing at an average annual rate of roughly 3·3 per cent *per capita* consumption in industrialized countries, and at the even higher rate of around 4·2 per cent in the developing countries.

Power, other than hydraulic, is typically obtained by converting heat into mechanical and, increasingly, then into electrical energy. Until recently the only real source of heat has been the combustion of fuel in air, but nuclear reactions now provide us with a new source. This is just as well, since the present world demand for power is of the order of 5770 million metric tons of coal equivalent per annum, and a fairly modest extrapolation indicates that this is likely to grow to around 30 000 million metric tons of coal equivalent per year by A.D. 2000. Even if world reserves of fossil fuels are adequate to meet demands of this order, the ecological consequences of burning fuels on such a scale must give rise to concern, and the satisfaction of the world's growing hunger for energy must therefore rely increasingly on the deployment of nuclear power. The fact that nuclear power is already economic in many places (depending on the local cost of alternative fuels) will merely accelerate this trend, and there is little doubt that nuclear is the prime source of energy for the long-term future.

The only route at present in sight for the utilization of this source of power is the processing of the heat produced by nuclear reactions into mechanical power by well-established thermodynamic cycles. For many reasons, any known process displays large economies of scale, and therefore the process will continue to be performed at very large central stations where the mechanical power is converted into electricity for distribution to the ultimate users over the existing electricity supply distribution networks.

Let us look first at the problem of their siting. The electrical utilities have long had the problem of where to build the large power stations which the constantly increasing demands for electricity require. The problem is a difficult one which need not be argued in detail here, except to say that the advent of nuclear power brings both new freedoms and new constraints into the situation. It brings new freedoms in that the nuclear power station has no problems of transport of enormous quantities of fossil fuel to the site or disposal of ash or very costly treatment of chimney effluents; but such a station does carry constraints which do not apply to other stations—there is the problem of nuclear safety, and in some cases problems of a lesser order, such as increased load on foundations and increased demand for cooling water. Of these, however, by far the most important is the question of nuclear safety with all that that entails.

Safety has always been a major consideration in the deployment of nuclear power. It is perhaps true that the early association of nuclear energy with atomic bombs and the fear of the relatively unknown hazards of ionizing radiation led initially to somewhat exaggerated public reactions to the potential dangers from nuclear installations, but it remains true that complete safety precautions are necessary in such installations. This fact is more keenly appreciated by the designer and owner of the plant than by most people; partly because he is in general very knowledgeable of the factors involved, but equally by his interest in the continued operation of the very expensive plant for which he has paid. No owner, even though he were free from Government control, would jeopardize his investment of scores of millions in a nuclear plant by accepting any risk of that plant proving inoperable owing to danger of any sort. Therefore, his economic incentives move him in the same direction as his public duty to ensure that no harm will ensue from any escape of significant amounts of radioactivity under normal operating conditions or indeed even under improbable accident conditions.

While therefore the owner has every economic (as well as legal) incentive to ensure the basic safety of the plant, all authorities throughout the world insist on further regulations to double-lock the door of safety against any nuclear hazard. Not only do they insist upon the most rigid examination of the double and treble safeguards built into the plant to prevent failure, but they also insist that if the totally unforeseen should happen and all precautions fail, then the consequential problem shall be capable of solution without real harm to the community. The precise mechanism by which this is achieved varies from country to country, but in general can be condensed into a requirement that nuclear stations shall be sited in an area sufficiently thinly populated to permit easy and rapid administrative action to be taken to safeguard the relatively few people affected.

It is this constraint of siting in thinly populated areas which is a main consideration in choosing sites for nuclear stations. It is undoubtedly a complicating factor—among other things it tends to force nuclear stations into the open country and, therefore, towards the more attractive parts of the countryside, but it is a constraint which utilities in the United Kingdom accept without question. The reasoning is that at the present relatively early stage of development of nuclear power, it is essential to be safe rather than sorry. Nuclear power is unquestionably the power of the future, and the welfare of humanity demands its rapid and economic development. Any nuclear accident with appreciable effect upon third parties can only have adverse effects on this development, and from this point of view as well as their normal responsibilities to the community at large, the electricity supply authorities accept the need of maximum precaution. However, there has been some progress in the United Kingdom. Initially very isolated sites were required but experience and development in designs have permitted some relaxation in siting closer to populated areas, thus giving greater freedom in the choice of sites and some reduction in the impact on amenity.

Apart from this requirement of relative isolation, the criteria for siting a nuclear station are, in principle, similar to those for other stations, that is, access to plentiful cooling water (which is the basic reason why so many stations are situated on estuaries or the sea coast, especially the early stations with their low steam temperatures and efficiencies and consequent greater demand for cooling water), good foundation conditions to carry the heavy weights of the plant, reasonable proximity to the electrical load to reduce transmission costs, and adequate and acceptable routes for the necessary transmission lines. Fuel delivery and spent fuel disposal are little problem since they involve transport of only a few tons per month, as against some millions of tons per year in the case of fossil-fuel-fired stations.

All nuclear stations so far built employ the same basic principle, namely, the heat from nuclear fission is converted into steam, which is then utilized to drive an orthodox steam turbine driving an electrical generator. The method of converting the nuclear heat into steam does, however, vary widely, as do the detailed design and fuelling arrangements of the reactors themselves, and it is interesting that the route chosen for the evolution of commercial nuclear power in various countries has, so far, largely evolved from whatever form their earliest experimental or prototype reactors took.

Thus the earliest reactors built in the United States of America, primarily for the production of military plutonium from natural uranium, were graphite-moderated and water-cooled, the heat in the fission process being wasted at low temperature. Before the heat produced could be used in the thermodynamic process it was necessary to increase its discharge temperature; for this to be done cooling water had to be pressurized to increase its boiling point to a useful temperature, the necessary pressures being of the order of 1000 lb/in² (70 bars) or more. Ordinary water is, however, an absorber of neutrons, and for convenience in design it is used not only for cooling but also as a moderator in place of graphite; natural uranium, however, has insufficient reactivity in these conditions to sustain a reaction, and enrichment of uranium was therefore necessary. This was in fact available to the Americans from their separation plant—again built for military purposes—and the pressurized water reactor can, at the expense of oversimplification, be regarded as a natural evolution of the original low-temperature plutonium-producing reactors. The hot water at high pressure is circulated through the reactor to a heat exchanger which produces steam used in orthodox turbines, the cooled water being returned to the reactor.

An alternative evolution was the boiling water reactor. This nuclear-wise is not dissimilar, but the water in the reactor is not only pressurized to raise it to a useful temperature but is also allowed to boil, the steam produced being used direct. In this case too the nuclear fuel requires enrichment.

Canada chose a different route. The original reactor, strictly for experimental purposes, although again built in connection with the military programme, utilized heavy water at atmospheric pressure for moderation and cooling. Heavy water absorbs far fewer neutrons than does ordinary water, and such a reactor can therefore be built utilizing natural uranium. The nuclear reaction proceeds more efficiently if the moderating heavy water can be kept cool; so the Canadian natural uranium heavy water power reactors use a cold moderator which surrounds pressure tubes in which the separate heavy water coolant is heated by the fuel. Because these tubes are within the reactor core they must be made of a material which does not absorb neutrons strongly, and zirconium is used for this purpose. This basic type of reactor which can employ natural uranium has been developed by varying the coolant, and proposals exist for combining boiling light water, a fog of light water or organic coolants with the basic heavy water moderated pressure-tube design.

The British followed a separate route. Their early plutonium-producing reactors at Windscale were graphite-moderated and air-cooled, partly because such a reactor avoided the neutron absorber of cooling water and could therefore be smaller than the water-cooled reactor, and partly because of the difficulties of direct river cooling in our comparatively small and densely populated country. From this a natural evolution was to enclose the entire reactor, change the cooling gas from air to CO_2 to avoid the oxidation of the graphite moderator, and to circulate the gas through a heat exchanger to produce steam. This, coupled with the use of a magnesium alloy ('Magnox' —hence the generic name) of low neutron absorption for cladding the nuclear fuel, permitted the use of natural uranium for power production, as in the Canadian case. The French followed a similar route.

The British, however, went further. They had available enriched fuel from the separation plant originally built for military purposes, and found that a more economic reactor could be built if slightly enriched fuel enclosed in stainless steel cans were used. This introduced neutron-absorbent steel into the reactor (necessitating enrichment), but permitted much higher temperatures to be used within the reactor, which in turn permitted the use of a more efficient thermodynamic cycle; the economic and thermodynamic gain thereby obtained more than paid for the additional cost of enrichment. This line of development is now being followed further and a new generation of gas-cooled reactors is under development, still gas-cooled and graphite-moderated, but using helium as the coolant gas and with the fuel in the form of particles of enriched uranium dioxide coated with an impermeable carbon coat and canned in graphite tubes. These fuel elements form part of the graphite moderator blocks which are themselves removed when refuelling and thus do not have to last the full 30-year life of the reactor.

There are other forms of power-producing reactor, for example the sodium-cooled graphite-moderated reactor and a helium-cooled version of the AGR of which prototypes have been built, but they have either not operated at all or have been unsuccessful, and many other concepts have never emerged from the conceptual stage. The vast majority of power reactors in use or being built are therefore of the pressurized or boiling water types, following the American concept; graphite-moderated, gas-cooled of the British; heavy water moderated and cooled, following the Canadian concept.

Attention in the field of reactor development is now moving towards fast breeder reactors. The main advantage of this system is that a much greater proportion of the fertile ^{238}U in natural uranium can be utilized and hence general fuel costs are much less sensitive to increases in the cost of uranium concentrate. It is not yet clear whether fast reactors are competitive with the thermal reactor types described above at current ore prices, but they will clearly have an advantage if ore prices rise significantly because supply cannot meet demand.

The point of major interest on all these reactors is 'which is the best?' and here there are as many opinions as there are designs.

It might be expected that the answer to 'which is the best?' would be fairly easy, and that all that is necessary is to take the price at which the buyer can purchase a station, settle the life over which that cost will be amortized, settle an expected reliability over that life, and hence calculate the amount of power generated during that life, add the fuel and other operating costs, and come up with a total cost per unit generated. The type of reactor which gives a minimum cost can then be taken as the best. Unfortunately things are not that simple, and it will perhaps be pertinent to discuss the four major components of the end result separately.

First the *capital cost of the station*, which typically will be many tens of millions of pounds. In a free competitive economy in which normal market forces have been given time to act, it can be assumed that the selling price has a reasonable relation to the seller's costs. This is not yet necessarily the case for nuclear power stations, for three reasons. First, the initial overhead costs of being in the business at all are massive. In order to enter this young and extremely complex technology, the firm has to engage a large number of highly skilled scientists and engineers and set out upon an expensive and lengthy programme of research and development. Allocation of an appropriate share of these overheads to the cost of any particular station clearly depends on a difficult judgement of the total market for such stations, and the share of that market which the seller expects to obtain.

Secondly, there are inherent and unavoidable risks in entering any new technological field, and the buyer for obvious reasons commonly requires the seller to guarantee that the plant to be bought shall be capable of meeting at least some specified requirements as to output, etc. The seller will obviously try to satisfy himself before quoting that the plant to be offered will be capable of meeting those requirements, but in any new technology there are risks and the seller must cover them by including in his price a contingency sum. The size of this contingency sum cannot be other than a matter of judgement and opinion.

Thirdly, the seller having decided to enter the business will obviously wish to maximize his share and may in fact choose deliberately to follow a loss-leader technique, thereby buying himself a large share of the market. His incentive to do so is increased by the large overheads he has to incur, and his desire to spread them over a maximum number of orders.

When added to these is the fact that the nuclear market is new and as yet comparatively small, and on both counts has not yet begun to stabilize, and further add that conditions of national prestige sometimes intervene, it will be seen that there are real reasons why the quoted price of a nuclear power station may have little relation to the true costs involved.

Facts of this type cannot be ignored by really large buyers. It is true that comparatively small buyers can sometimes step into the bargain basement and buy a station at a price which, whatever the economic results to the seller or indeed to the selling country may be, is a bargain price to him; but a really large long-term buyer cannot take this attitude. The reason is that long-term sellers are in business to make a profit which they can only obtain from the buyers, and therefore, long-term, the price to the buyer must be related to the true costs of the seller. This is a matter of major import. It may well pay the seller to encourage, by loss-leaders or other ways, a shape of technology which is not the long-term optimum but to which he has committed himself; if the buyer concurs, then long-term his costs are higher than they need be. The buyer's essential costs will only be minimized if he chooses what is fundamentally the 'best', which may not always be the 'cheapest' type of station.

The second main head of costs concerns the *availability of the plant in operation*. All forms of nuclear power station have low or very low fuel costs, and therefore can be expected to be run very nearly at full load whenever possible throughout their lives. However, like all plant, nuclear stations are liable to breakdown and also have to be overhauled, and the degree of susceptibility to breakdown, or the complexity, time taken, and cost involved in routine maintenance of the differing types are questions of technological judgement, but in this new field such judgement has to be based on comparatively little hard experience. Nevertheless, availability is almost as important as length of life in calculating total costs, and has a wide bracket of justifiable assumptions.

Thirdly there is the question of the *life over which the capital costs are to be amortized*. Technically this is a matter of severe technological judgement, but again such judgement has to be based on comparatively little hard experience, and in a project as capital intensive as is nuclear power, the judgement is of great importance in calculating the final costs of generation of each unit. Further, the technological factors involved in the judgement may be—commonly are—quite different in the different types of station on offer, so that when comparing differing types of stations no sweeping generalizations on 'proper' lives are really tenable. It is a fact that quite responsible assessments of nuclear costs from a power station of given type have been prepared based on lives as low as 15 years on the one hand and as high as 40 years on the other; as a result total costs per unit from basically similar plant vary considerably.

Finally, there is the question of *fuel and other costs*. These are probably the most exactly calculable of all costs, provided always that reasonable assumptions are made as to the future cost of uranium and any enrichment it may require for the type of station under consideration. Fuel costs are, however, important. Reactors using natural uranium are inherently large, rather difficult to build, and have a comparatively low heat-output per unit volume. It follows that they are expensive in first cost and in practice their running costs have not turned out to be significantly lower than those of the better enriched uranium reactors. The burn-up which can be obtained from natural uranium is limited by reactivity considerations and the advantages of higher burn-up which enriched uranium gives may well outweigh the cost of the enrichment process.

While fuel costs are thus capable of relatively precise calculation compared to the other three main factors described above, the fuel design and enrichment is so intimately linked to the design (and therefore cost) of the reactor itself that fuel cost alone forms no basis of comparison between reactors.

In spite of all these uncertainties, comparisons of the merits of different reactors have to be made, and are of very considerable importance since tens or hundreds of millions of pounds are to be staked on the judgement. Let us therefore ignore the uncertainties, and consider how a judgement should be made on the assumption that the basic figures of prices are properly related to true costs. This is the question of the so-called 'ground rules' for calculation of the cost of the electricity produced; in other words, how do we treat the four main heads of costs enumerated above?

First again is the *capital cost of the station*. This can be divided into four main sub-headings, each with considerable capacity for different presentation. These sub-headings are as follows.

(a) The main contracts—broadly the plant and buildings. These are the largest items, but are not necessarily totally defined in a tender price. Probably few nuclear stations have been built to the price tendered for this *tranche* of a quotation, sometimes because the extrapolations inevitably made at this early stage of the art have proved unobtainable (or perhaps unwise) by the time postulated, sometimes because clearly desirable improvements become possible as development proceeds and the costs of the job are adjusted according-ly, and sometimes because of straight engineering or site difficulties or similar reasons; inflation during the long construction period can also be relevant. All that need be said on this aspect is that a buyer should make as certain as possible that the indeterminates inherent in the tender (and there will be some) are reduced to a minimum, and then be covered by a suitable contingency sum in his own estimate.

(b) Other contracts. These include land, site levelling, drainage, cost of access both on and off the site for delivery of very heavy loads, circulating water-works, landscaping, fees, engineering, etc. In aggregate these can amount to somewhere around £5 per kW, which is significant when comparing the relative merit of competing tech-niques.

(c) Interest during construction is a charge on the buyer, but may or may not be presented in a statement of capital costs. Dependent on the actual incidence of payments, discount rate, and total construction period, interest can amount to a really significant figure—at 8 per cent discount perhaps equivalent, on a 'present worth' basis referred to the mid-date of commissioning, to well over 20 per cent of the construction costs.

(d) Initial fuel charge is usually considered, in the United Kingdom at least, as a capital item, and covers initial fuel, some spare components, and some interest during manufacture; this again may or may not be included in a statement of capital cost for the station. The cost of initial fuel can be treated in different ways for accounting purposes, ranging from a revenue cost spread over the life of that first fuel charge, to a capital cost amortized over the life of the station with replacement fuel only debited as a revenue cost.

There is also the question of inflation during construction. British practice is to express costs in present-day money values and to meet inflation by prearranged contract price adjustments; in United States practice it is more usual to include a sum to anticipate inflation.

The sum of the costs in (a), (b), (c), and (d) is the total capital cost, but when comparing techniques the comparison does not end there. As well as the inherent variables listed above, there must be considered:

(i) The size of the unit. Most nuclear power stations show considerable economy of scale—roughly at the rate of rather over 15 per cent reduction in specific capital cost for a doubling of unit size.

(ii) The number of identical units forming the station. Taking the specific cost of a two-unit station as 100 per cent, that of a single unit station will be roughly 118 per cent and that of a four-unit station about 93 per cent.

(iii) The site conditions, for example relative ease and cheapness of cooling water supply, foundation conditions, etc.

(iv) The extent of supply, including the electrical transmission connection.

Turn now to *availability* over the life of the station—a most important factor in any cost study, but still quite largely a matter of opinion reflecting both technical judgement and the degree of financial conservatism adopted. It is commonly assumed that in the early months teething troubles may cause lower availability (in this respect the early Magnox stations were a pleasant surprise) and that increas-

4

ing the size of the units may tend to decrease availability. The question is important, for the loss of one percentage point in lifetime availability may be worth about £1 per kW present worth. The C.E.G.B. at present assume an availability of 75 per cent in each year of life; the utilities in the United States tend to adopt 80 per cent or even 85 per cent.

The life given to the station is equally a summation of technical judgement and severity of financial policy. The C.E.G.B. currently assume, for the purpose of economic comparison, a life of 20 years for Magnox and 25 years for later stations although all components are designed with a target life of 30 years; in the United States longer lives are usually assumed—30 or 35 years.

Fuel costs over the life of the station depend basically on ore costs, enrichment costs, and fabrication. Ore (or more correctly, uranium ore concentrate, UOC or U_3O_8) is currently priced at \$7–\$8 per lb. Fairly recently it was thought that cheap resources would be exhausted in a few years and that prices would rise, but prospects of holding them now appear better. Enrichment costs vary—the plant involved has large economies of scale but in both the United Kingdom and the United States they were originally built for defence purposes— and, further, the cost is dependent on the price of electricity, which is cheaper in the States (where coal costs about 1p per therm) than in the United Kingdom (coal 2·2p per therm). The advent of the centrifuge method of enrichment which is now being exploited collaboratively by Britain, Germany, and Holland may change this latter situation. The method requires far less electricity than the gaseous diffusion process used in the earlier plants so that the inherent advantage given by cheap American coal will disappear. On the other hand, very large scale manufacture of the centrifuges themselves is necessary if their capital cost is to fall to a value which will make the process cheaper than the diffusion process based on low-price coal. Fabrication costs differ widely with the actual design and materials used for fuel and cladding: typical British figures are shown in Table 3.1.

TABLE 3.1
Typical fuel costs for British nuclear power stations

Reactor Type	Enrichment	Rating (MW(th)/t)†	Burn-up (MW(th)/t)	Fabrication (£/tU)	Uranium (£1/t as UF_6)
Magnox	Natural 0·7	3	3 600	12 500	
AGR	2·2	13	18 000	30 000	70 000
HTR	6·0	50	60 000+	150 000	250 000

† MW(th)/t = megawatts (thermal) per tonne.

It will be appreciated from the above discussion that the possible (and entirely legitimate) variations in ground rules can make quite large differences to the outcome of the calculation of costs of nuclear generation. Such calculations are not worth even looking at unless the ground rules adopted are known and preferably stated on the results sheet. Once settled, these same ground rules can be applied in principle to either coal- or oil-fired stations to give a first indication— and only a first indication, the point will be referred to again—of the comparative merit of nuclear as against fossil-fuel-fired stations.

Below are listed the results of such exercises for some British stations. The ground rules are as follows:

(1) The construction costs
   (a) include the cost and capacity of emergency gas turbines;
   (b) exclude the initial fuel charge;
   (c) exclude interest during construction.

(2) The generation costs have been calculated on the basis of capital charges costed on
   (a) an annuity basis appropriate to an 8 per cent interest rate;
   (b) the capital costs including interest during construction on an 8 per cent basis;
   (c) the capital costs including the initial fuel charge;
   (d) a life of 30 years for coal- or oil-fired stations, 25 years for AGR stations, 20 years for Magnox stations;
   (e) an assumed lifetime load factor of 75 per cent;
   (f) excluding the cost and output of emergency gas turbines (since these are also used for peak load as well as emergency purposes they provide useful capacity on the C.E.G.B. system and may be regarded for cost purposes as a separate station);
   (g) including a royalty payment for AGR stations.

Construction and generation costs for some British nuclear stations are given in Table 3.2. For purposes of comparison, the costs of roughly contemporary fossil-fuel-fired stations are given in Table 3.3.

Table 3.3 shows some interesting features:

(a) The rise of capital costs with inflation in a nearly static technology. This effect is more than concealed in the nuclear case with its advancing technology.

(b) The quite significant variation of total costs due to factors other than capital costs. This of course reflects the variations of delivered fuel cost at the different sites. Since all but the last of the nuclear

TABLE 3.2

Construction and generation costs for some British nuclear stations

Station	Year commissioned	Number and capacity of reactors (MW(E))	Construction cost (£/kW so)†	Capital charges (p/kWh so)	Running costs (p/kWh so)	Total costs (p/kWh so)
Berkeley	1962	2 × 138	185	0·40	0·15	0·55
Bradwell	1962	2 × 150	175	0·38	0·13	0·51
Hinkley Point A	1965	2 × 250	154	0·38	0·13	0·51
Trawsfynydd	1965	2 × 250	146	0·31	0·11	0·42
Dungeness A	1965	2 × 275	117	0·25	0·10	0·35
Sizewell A	1966	2 × 290	107	0·23	0·10	0·33
Oldbury	1967	2 × 300	113	0·25	0·09	0·34
Wylfa	1970	2 × 590	111	0·25	0·09	0·34
Hinkley Point B	1972	2 × 660	76	0·15	0·07	0·22
Hartlepool	1974	2 × 660	71	0·15	0·07	0·22

The first seven stations are Magnox type in full commission; the eighth is Magnox and at time of writing being commissioned; the ninth and tenth stations are AGRs under construction. The downward trend of costs with time, in spite of inflation, is unmistakable.

† kW so: kilowatts sent out.

TABLE 3.3

Construction and generation costs for some British fossil-fuel-fired stations

Station	Year commissioned	Number and capacity of units (MW)	Construction costs (£/kW so)	Capital charges (p/kWh so)	Running costs (p/kWh so)	Total costs (p/kWh so)
Coal-fired—Midlands and Northern England						
Blyth B	1962	2 × 275, 2 × 350	41	0·06	0·24	0·30
Drakelow C	1965	2 × 350, 2 × 375	45	0·08	0·20	0·28
Ferrybridge C	1966	4 × 500	40	0·06	0·19	0·25
West Burton	1967	4 × 500	42	0·07	0·20	0·27
Ratcliffe	1968	4 × 500	42	0·07	0·20	0·27
Drax	1972	6 × 660	55	0·09	0·20	0·29
Coal-fired—Southern England and South Wales						
West Thurrock	1962	2 × 200, 3 × 300	47	0·08	0·27	0·35
Tilbury B	1968	4 × 350	49	0·09	0·26	0·35
Aberthaw B	1970	3 × 500	49	0·08	0·26	0·34
Oil-fired—taxed oil						
Pembroke	1970	4 × 500	48	0·08	0·18	0·26

stations listed in Table 3.2 are sited where fossil-fuel costs would be high their generation costs should be compared with Tilbury and Didcot, but Hartlepool should be compared with Pembroke, noting however that the dates of commissioning are not the same.

The trends of capital costs with time are worth plotting for the two competing technologies—nuclear and coal. This is done in Fig. 3.1, and brings out the sharp reduction in capital costs with advancing technology in spite of inflation in the nuclear case, as against the battle with inflation fought in the coal case from the 1950s through to the mid-1960s, won by advancing technology until then, but there-after—so far at least—a losing battle.

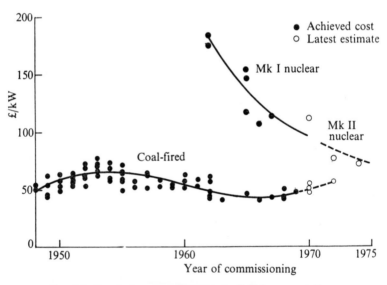

FIG. 3.1. Trend of construction costs for British power stations.

Doubt is sometimes cast—as indeed has been done in this chapter— on the accuracy of some estimates of nuclear costs. It is therefore worth recording in Fig. 3.2 below, the British history on this so far. The effect of changing money values during the construction period of a station is excluded from the achieved cost shown on Fig. 3.2, i.e. no adjustment has been made for inflation.

Turning now to nuclear stations other than British, it is more

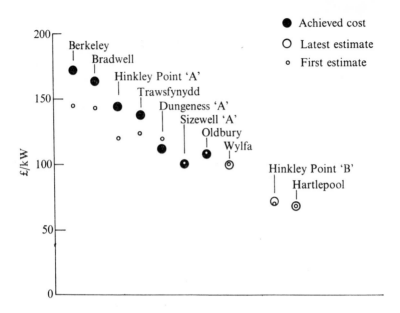

FIG. 3.2. Accuracy of estimates of construction costs for British power stations.

difficult to establish facts on bases which are known to be strictly comparable. Table 3.4 does, however, reproduce a selection from a tabulation published in *Nuclear Engineering International* in February 1970, and quoted here by permission.

Table 3.4 brings out immediately the extreme difficulty of making any sort of comparison of either capital or generation costs without firm knowledge of the ground rules on which they are based and indeed the circumstances in which the price was fixed. But even accepting that there is no indication of the extent of supply, the treatment of interest during construction, the treatment of initial fuel cost and the other variables already described, the identifiable variations in the table are interesting. Thus the scatter of capital costs in one technology—the PWR—at a capacity of about 850 MW is quoted as between £57 and £107 per kW; the load factors assumed are mostly 80 per cent, but one is as high as 88 per cent average over a 30-year life; the assumed lives vary from 25 in the lowest case to 40 years for the two highest, and the resultant generation costs vary from 0·08p/kWh for a BWR commissioned in 1969 to 0·27p/kWh for

TABLE 3.4

Costs of some water reactor stations
(Costs converted at the rate of $2.40 to the £)

Station	Type	Output (MW)	On power (date)	Capital cost (£/kW so†)	Life (years)	Load factor (%)	Interest rate (%)	Generation cost (p/kWh)
Haddam Neck	PWR	600	Jan. 68	66	25	80	5	0·20
San Onofre	PWR	450	Jan. 68	79	30	84	—	0·17
Oyster Creek	BWR	640	Jun. 69	56	30	88	7¼	0·08
Vermont Yankee	BWR	537	Jul. 71	97	30	—	—	0·27
Fort St. Vrain	HTR	330	72	66	—	80	—	0·20
Brown's Ferry	BWR	3 × 1098	Oct. 71	59	35	85	—	—
Cooper	BWR	778	Apr. 72	75	35	80	—	0·19
Maine Yankee	PWR	830	May 72	90	28	—	—	0·24
Indian Point	PWR	873	Apr. 76	57	30	80	11¼	0·17
Kewaunee	PWR	527	Mar. 72	89	40	80	7¼	0·08
Arkansas 1	PWR	840	Dec. 72	70	40	85	9	0·17
Beaver Valley	PWR	847	Jun. 73	107	32	85	6¼	0·20
Malibu	PWR	490	Jun. 75	80	28	80	4	0·20

† kW so = kilowatts sent out.

the same technique in a unit of comparable size commissioned two years later—a spread of more than 300 per cent. It is difficult to do more than draw attention to the variation in the figures and to comment that in so far as they are lower than the costs quoted for the British cases, this is largely due to the longer life and higher load factors assumed, and perhaps express doubt as to how other costs included in the British case have been treated.

Lastly on this question of comparing technologies, the straight comparison of costs per unit generated at an assumed load factor over an assumed life, while useful in giving an indication of relative merit, is grossly inadequate as a means of choosing the best investment. The reason is that the characteristics of the costs in, for example, the nuclear versus the coal case differ; nuclear power tends to have high capital costs and low running costs. For this and for other reasons the load factor on a coal-fired station on a growing system will certainly fall faster than on a contemporary nuclear station, and this fact will alter the consequential load—and costs—of all other stations on that system. The problem of exact calculation of the relative merit of nuclear as against an alternative type of station is therefore extremely complex and really involves a calculation of the total costs of the entire system, first with the addition of the nuclear station and then again with the alternative addition of the coal station, over the projected life of the competing stations. Such a calculation is not only difficult but is also special to the particular generating system concerned, and thus is of little interest outside that system. It is perhaps for this reason that the relatively simple and widely accepted basis of equal load factors over the entire station life is used for comparison of, for example, nuclear against coal, but it is a basis which, in fact, should be regarded only as a guide.

There is one further source of information which is certainly worth quoting. That is a report to the Joint Committee on Atomic Energy of the Congress of the U.S.A. on 'Developments in nuclear power economics—January 1968 to December 1969' by Dr. Philip Sporn, and here quoted by permission of the author. The report—authoritative as are all papers from Dr. Sporn—repays study by anyone interested in energy and its development, and to pick out of its context his assessment of nuclear power costs does it less than justice. The figures he gives are priced at 1 July 1969 for a nuclear station to be completed in 1976:

Capital cost—1100 MW unit	203·5 $/kW (£84·8)
Load factor	80%
	mills/kWh (1mil = 0·1 cent)
Fixed charges at 14 per cent	4·07
Fuel charges	1·70
Operating and maintenance	0·30
Insurance	0·10
Total costs at switchboard	6·17 mills/kWh (0·261p)

Dr. Sporn uses 14 per cent as the fixed annual charges on capital. For the purpose of estimating costs of generation, the C.E.G.B. use an annuity based on 8 per cent interest rate and for 25-year life this gives 9·4 per cent annual charges on capital. (For the separate purpose of investment appraisal the C.E.G.B., in accordance with a Government request, use 10 per cent interest rate, which gives about 11 per cent capital charges.)

Since Sporn's paper is largely concerned with the recent apparent retrogression of the comparative economics of nuclear power in the United States, it is worth while to quote his estimate of the costs given by a contemporary coal-fired station in the States. In such comparisons the price of coal is of course all important, and Sporn takes this at 2·5 cents (1p) per therm. His figures then become:

Capital cost—800 MW coal-fired unit	170·6 $/kW (£71·1)
Load factor	80%
	mills/kWh
Fixed charges at 14 per cent	3·41
Fuel charges	2·19
Operating and maintenance	0·30
Insurance	—
Total costs at switchboard	5·90 mills/kWh (0·246p)

and he deduces that, on these bases, nuclear power becomes competitive if coal costs more than 2·8 to 3·0 cents per therm. This conclusion is interesting, particularly bearing in mind that the average delivered cost of coal at British power stations was in March 1970 over 5 cents (2p) per therm.

What then is the general conclusion to be reached from this confusing welter of figures? Surely, it must be that even if the relative merits of the competing systems of nuclear generation are not as clear as their individual proponents would wish us to believe, nuclear power is already competitive in high fuel-cost areas. Further, it is a young and rapidly developing technology and therefore its real costs can be expected to fall as techniques progress. On the other hand, the technology of coal-fired stations after many decades of intensive development is for the moment relatively static, and the prospects of really significant reductions in the cost of winning coal are not immediately obvious. Add to this the continuous escalation in power demand and the increasingly costly problem of controlling air pollution from the burning of fuel, and there seems little doubt that nuclear stations must make the major contribution to future power supplies.

The sociological consequences will be large. On a national level, as an economic alternative to coal, it has and will continue to contribute to the contraction of the coal industry, and thereby add to the problems faced by that industry and indeed the nation in redeploying man-power in areas where alternative employment is usually scarce. The fall in coal consumption must also reduce the demand for transport; so British Railways will be affected. Power stations need no longer be sited close to coalfields or indeed to oil ports, and this new flexibility means that such stations are not in principle confined to locations in or close to the traditional industrial areas. The effect of these very large works on non-industrial areas can be profound. They can, unless great care is taken, affect amenities, but they carry with them attendant local economic advantages. The Central Electricity Generating Board sponsored a survey carried out by the University of Aberystwyth on the local social and economic effects of building a large nuclear station in a relatively poor rural county, which showed that some of these benefits are felt immediately—more money spent in local shops, more money spent on roads and other communications, increased pressure to improve local educational facilities for workers' children, and so on; longer-term results are increased local revenue from payment of rates and a new source of permanent employment. On the social side the Board and its staff give every encouragement to the integration of the station and its workers into local life, and both appear to have settled down as an integral part of the local community very quickly.

In short, nuclear power is a development altogether timely at this stage of the world's progress, and almost wholly beneficial in its effects.

The potentialities of the development of nuclear energy for purposes other than warfare and power

By H. Kronberger, C.B.E., F.R.S.

Let us, in this last technical chapter, remind ourselves of the simple process of fission on which the wide spectrum of nuclear technology and its future potential is based.

A neutron enters the 'fissile' atom (uranium or plutonium) and produces in it an internal upheaval which causes the atom to split violently into two parts, roughly half the size of the original atom (and these fission fragments usually undergo further radioactive transmutations); at the same time two or three neutrons are emitted, of which (in a controlled and self-sustaining fission reaction) one is used to produce a further fission while the others are absorbed in the surrounding matter; in addition, each fission is accompanied by a flash of high-energy gamma radiation—electromagnetic radiation like light, but of much greater penetrating power. As these highly energetic products of fission are brought to rest in the surrounding matter, their energy appears as heat; they can also convert some of the surrounding matter into unstable forms (radioactive isotopes) which, like the fission products, can themselves emit particles and high-energy electromagnetic radiation.

Thus nuclear technology has three aspects: the production of heat, of radioactive isotopes, and of radiation. Accordingly the scheme of this chapter is as follows. First I shall describe uses (other than those covered in the previous chapters) to which the heat from the fission process can be put in explosions or in reactors; then, uses of radioactive isotopes; and finally the applications of nuclear and radiation physics, using as typical examples those related to the exploitation of natural resources. Within the framework of a single chapter it is of course impossible to enumerate all the potential applications of this vast technology; rather is it intended to indicate that the impact of the discovery of fission has much wider implications, and can have much more immediate benefits, than the making of weapons and of power stations.

Heat from nuclear fission

Peaceful nuclear explosions (PNE)

In terms of the cost of the energy released, nuclear devices are the cheapest form of explosive, and in principle they should make it possible to move earth and rock economically on an unprecedented scale, and to carry out civil engineering operations which would otherwise be impracticable. For instance, a 100-kiloton nuclear device (corresponding approximately to 100 000 tons of TNT) which would cost about £200 000, would yield a crater 320 ft deep and 1200 ft in diameter, and 50 million tons of broken rock—all in a single shot!

There are two types of application envisaged for PNEs: cratering explosions and deep underground explosions. In cratering explosions the device is detonated at shallow depth; a series of craters can be used to form a cutting or a canal. This type of rock-moving operation opens up the possibility of building canals, harbours, and large water-reservoirs; the diversion of rivers; and the removal of overburden for open-pit mining. Such projects if carried out by conventional means would be either impossible or of prohibitive cost.

The deep underground explosion produces a large spherical cavity surrounded by shattered rock. Deep explosions can be used for making underground caverns for gas storage, for easing or stimulating the flow of gas or oil from surrounding regions, or simply for fracturing the rock to ease the extraction of minerals.

The possibilities are tremendous—but how safe are such explosions? In shallow cratering explosions there are three main hazards: radioactive fall-out, air blast, and ground motion. Most of the radioactivity is trapped within the material which falls back into the crater, but its long-term future (taking into account the effect of leaching out by water) must give rise to concern. The radioactivity which escapes, and of which some 90 per cent is deposited within six miles of the detonation, can be minimized by thermonuclear explosives having very low fission yield.

The major radioactive hazard is then from tritium which gives rise to tritiated water. Air blast will cause direct damage near the detonation, but under unfavourable atmospheric conditions window breakage can occur at distances of more than 50 miles. Ground motion can cause considerable architectural damage within say 10 miles. It is most unlikely that shallow nuclear civil engineering could be carried out near populated areas.

In deep underground explosions air blast does not occur, and there is no prompt release of activity into the air; about 90 per cent of the activity is trapped in the form of an insoluble glass formed from the molten rock. The main hazard is then the contamination of stored or extracted products (such as natural gas) by tritium, and of ground water by radioactive material migrating through rock strata. Ground motion is still important and the site would have to be more than 10 miles from populated areas.

Because the safety problems of deep underground excavations are less severe, industrial interest in the U.S.A. and U.S.S.R. tends to concentrate on the application of contained underground explosions in stimulating the flow of oil and gas. In some measure the experiments have been successful: the major question is how to use the slightly tritiated gas without creating a radiation hazard.

Thus at present no single industrial application of nuclear explosive engineering has been demonstrated as safe, but its potential is considered so great that during the negotiation of the Non-Proliferation Treaty, which came into force in March 1970, the non-nuclear weapons states insisted that their PNE interests should be protected and the nuclear weapon states promised to make available the benefits of peaceful nuclear explosions to those non-nuclear weapons states which were party to the treaty.

Nuclear desalination

As distinct from the nuclear explosion, nuclear heat in a reactor is released in a slow and regulated manner, suitable for conversion into mechanical or electrical energy in a heat engine such as a steam driven turbogenerator. A heat engine, whether it derives its heat from a reactor or from fossil fuel, converts a fraction (say 40 per cent) of its heat into useful mechanical or electrical energy—the rest of the heat is rejected at the lowest possible temperature into the cooling water, i.e. the sea or a river.

The importance of nuclear power to desalination (the production of fresh water from sea or brackish water) arises entirely from the fact that in many countries very large nuclear power stations are, or will be, the cheapest source of heat and of electricity.

Desalination processes fall into two categories: those which require their energy in the form of heat—such as the distillation of sea-water —and those which use mechanical or electrical energy. In the freezing process, where salt-free ice is produced by freezing sea-water, the energy is needed to drive the compressors of the refrigerators; in reverse osmosis, pumps are required to force the brackish water through cellulose acetate membranes which allow the water, but not the salt, to pass through; in electrodialysis, direct current electricity is used to pull the electrically charged components of the dissolved salts (the ions) out of the solution.

Desalination must be considered as a manufacturing or production process: it requires capital equipment and energy. The problem is that in order to compete with the natural product, the cost of 'manu-factured' water must be low—for industrial and domestic purposes (other than in exceptional circumstances) it should not exceed around 5p per tonne; for agricultural water (irrigation) the price would have to be a good deal lower. I cannot think of a manufactured product of any kind which could be made for under 5p per tonne other than desalinated water, and that only from desalination plants using heat or electricity from large nuclear power stations.

The sea-water distillation process is now commercially well established, in units producing a few million gallons of water (i.e. 5000 tonnes and over) per day at a cost of perhaps 50p per 1000 gallons (say 10 to 15p per tonne). Such plants are fuelled by gas or oil. Many communities in arid areas (for instances at the oilfields in the Arabian Gulf) now depend on such desalinated water; the cost of the water is high, but that of the alternatives (such as shipping in the water) would be even higher.

When considering nuclear heat for large distillation plants, the scheme which is economically the most attractive is the 'dual purpose plant'. Steam at high pressure and temperature is raised in a reactor; as it does its work in passing through the turbine to generate electricity its pressure and temperature falls; when the steam has reached a temperature suitable for heating the sea-water in a distillation plant (around 125 °C) it is passed to the distillation plant instead of driving further stages of the turbine. In this way about three-quarters of the useful energy content of the steam, i.e. that part of the heat content capable of conversion into useful mechanical or electrical energy, will have been used to produce electricity; the rest of the useful energy is sacrificed to provide the heat needed in the distillation plant. After allowing for the loss of electricity production this 'pass out' steam is probably the cheapest form of nuclear heat for sea-water distillation, and, incidentally, for district heating and for many other industrial processes requiring heat at modest temperature.

However, in order to get down to water costs of around 5p per tonne the plant has to be large: a typical dual purpose scheme would be one producing 100 million gallons (450 000 tonnes) of water per day, together with perhaps 400 000 kW of electrical output; the water and electricity demand should ideally remain matched during the seasons. Such schemes can be considered for industrialized countries, but very few developing countries could use this amount of electricity, tied to a single unit. Nevertheless such projects are intensively studied and, as paper plans, taken to enormous proportions in the form of 'nuclear agro-industrial complexes'. In these, huge nuclear power stations are envisaged, providing electricity for energy-intensive industries such as fertilizer or aluminium production, water for the industry and population associated with them, and it is hoped water cheap enough to be used for agriculture. Technically such schemes present few difficulties: reactors and desalination plants of the required size could be built on present-day technology. The problems are those of the dependence of a large agro-industrial community on the reliability of one or two huge power plants, of administration and above all, of finance.

I was once described as a sceptic when in an after-dinner speech I facetiously remarked that the problems of the 'agro-industrial complex' would lead a financier to seek the services of a psychoanalyst rather than of a technologist. However it is only about the timing of such schemes that I am sceptical; at the moment there are too many other demands for the finance which such schemes would require.

Desalination methods depending on energy, rather than on heat, are perhaps more likely to benefit immediately from cheap nuclear power: they need not be located at the power station but can take electricity from a grid supply system; their costs do not increase as sharply with reducing unit size as nuclear stations, and units could be built at several locations around the country. It is my belief that in this more modest form desalination, deriving benefit from cheap nuclear power and possibly from preferential electricity tariffs, will be introduced in the U.K. fairly soon to augment natural water supplies. Under conditions of rapidly increasing water demand, the alternative is building another reservoir, to give a firm supply for, say, the next 20 years; this means that the reservoir will be under-utilized for the early part of its existence, and capital will have been spent too early. With the present high interest rates it may be cheaper in certain locations to match the increasing demand by the successive installation of desalination plants.

Other uses of fission-generated heat and electricity

Desalination is one instance where heat or electricity is used to carry out a production process. Chemical and metallurgical production processes are conveniently divided into thermal and electrolytic. In the former the reaction is promoted by raising the temperature; in the latter the reaction is carried out electrically by means of the transfer of electric charges. The blast furnace is an example of the former; the aluminium smelter of the latter. In many cases thermal and electrolytic processes are alternatives and the choice between them may then depend on the relative cost of electricity and of heat plus reagents to carry out the thermal reaction. As nuclear power progressively reduces the price of electricity, so more and more processes will turn to electrochemistry.

Thermal processes operating at modest temperature can derive benefit from nuclear power through cheap 'pass out steam' as described earlier. But many chemical, and in particular metallurgical, processes require much higher temperatures in the region of 800–1000 °C. This temperature is still above that of the high-temperature reactors which are being developed today; nevertheless studies are already being carried out on the possible use of high-temperature reactor heat in the steel-making industry. However, it will necessarily be a long time before the feasibility and economic merits of such schemes can be fully established.

Radioactive isotopes and radiation

The uses of radioactive isotopes and their radiations are so manifold that the following paragraphs can give only an indication of the field. In fact it would be very difficult to list all the uses to which isotopes have been put in scientific research, medical diagnostics, therapeutics, agriculture, and industry.

5

Radioisotopes are unavoidable by-products of the operation of nuclear reactors, but they have proved to be so valuable that their production is now an important function in itself. Radioisotopes of a wide range of elements accumulate as fission products in nuclear fuel, from which they can later be separated during chemical reprocessing, and they are also formed when foreign materials are bombarded with neutrons in or near the core of a reactor.

Radioisotopes give off characteristic radiations which are readily detectable and which may be of great penetrating power, and each radioisotope has its own decay-rate, expressed as 'half-life' in which the initial activity of any quantity drops by half. The radiations interact with matter in specific and often useful ways. Apart from these specifically nuclear qualities, the radioisotopes of any element are virtually indistinguishable chemically, physically, or biochemically from the normal (inactive) isotopes of that element. These characteristics enable radioisotopes to be used as sources of signals in a range of industrial measuring instruments, or as 'tracers' for studying the movements and distribution of materials in many kinds of physical, chemical, or biological systems. They are also used as energy-sources for bringing about certain changes (particularly in living matter), or for providing limited amounts of more directly useful energy. Their applications, which have by no means reached their full potential for exploitation, extend into almost all fields of scientific research, into most branches of industry and public works, into agriculture, and (above all) into medicine.

Hazards of radioactivity

The major limitation on the use of radioactivity lies in its inherent biological hazard; the radiations have the power of causing serious or even fatal latent damage to living organisms of all kinds, without at the same time giving any of the usual warnings through the senses. The damage can be cumulative, and may not become apparent to the victim for years; in the case of genetic damage it may come to light only in a later generation. For this reason the use of radioactive materials is everywhere subject to strict legislative controls which are based on a large measure of international expertise and agreement. As a result the dangers of radioactivity have themselves proved to be a most effective assurance of radiological safety, and deaths or injuries arising directly from the practical uses of radioisotopes have in fact been extremely rare.

Medical uses

The first significant practical use of radioactivity was in the medical field, when radium was found to possess certain curative properties. Because of its novelty its use was quite widely advocated before its attendant hazards were fully appreciated, so it was perhaps fortunate that its extremely high cost (reaching £8 per milligramme at one time) imposed a severe practical limitation on its use. By the time artificial radioisotopes became available, at very much lower effective costs, the dangers were much more fully understood, and a more cautious attitude was adopted towards the use of radioactivity in any form. At the same time new and more effective therapeutic methods were developed, and a whole range of diagnostic procedures based on 'tracer' techniques was evolved.

The power of ionizing radiations to kill or severely damage living tissues makes them a valuable supplement to, or substitute for, surgical methods, and they have proved especially valuable in cancer treatment. Deep-ray therapy units based on cobalt-60 are in successful use in hospitals throughout the world, and show a substantial cost advantage over X-ray installations of equivalent effectiveness. Some deep-seated tumours are better treated by implant or perfusions of suitable short-lived radioisotopes (e.g. gold-198) at the site of the trouble, and a similar method is used to destroy the pituitary gland with yttrium-90 implants without drastic surgery. Certain types of skin cancer or fungal infection, and some conditions of the cornea, are treated by direct external application of phosphorus-32 or strontium-90 ointments or applicators. Blood conditions such as polycythaemia vera can be treated by injection of phosphorus-32 as phosphate. Some neoplasms of the thyroid, or over-activity of the gland itself, can be treated with iodine-131, using the body's own mechanism of iodine uptake to convey the isotope to the site.

Tracer diagnostic techniques are based on the virtual identity of the chemical or biological behaviour of a radioisotope with that of its inactive counterpart, combined with its continued emission of readily detectable, and often highly penetrating, radiation. Food components, drugs, chemicals, and even constituents of the body-fluids can be 'labelled' with suitable radioisotopes, and their subsequent movements followed quantitatively and in considerable detail in the body by external detectors, or by measurements on biological samples. One of the first such procedures to be perfected, and probably still the one that is most widely used, is the diagnosis of thyroid malfunction by iodine-131, sometimes as a preliminary to the thyroid treatment described above. Radioactive iodide solution is administered orally, and tends to concentrate at the thyroid at a rate dependent largely on the organ's health; the build-up of radioactivity is detected externally, and automatically recorded to give a growth curve of immediate diagnostic significance. By an extension of the same technique, the location, shape, and size of the gland (and of any associated neoplastic growths) can be accurately determined. Other radioisotopes are used to study such conditions as blood circulation, heart-lung function, vitamin uptake, renal function, plasma volume, red-cell population, internal haemorrhage, or brain-tumour location, as well as being used for many research investigations on human volunteers or laboratory animals, or *in vitro*. Radiation therapy and radioisotope diagnostics have become thoroughly established in present-day medical practice and research, and represent one of the major contributions of physics to medicine.

Agriculture

Radioisotopes are widely used as tracers in agricultural research, and to a much lesser extent (at present) as radiation sources. Major tracer applications include studies of plant nutrition, and cover the uptake, metabolism, and distribution of nutrient and tracer elements from the soil or from added fertilizers, using appropriate radioisotopes and a range of electronic and photographic measurement techniques. Perhaps the most striking research success has been the elucidation of the all-important process of photosynthesis, achieved by growing plants in carbon dioxide containing carbon-14 and studying the build-up and breakdown of radioactive products under the action of light and chlorophyll.

Equally important studies have been made on insects, both to add to the general understanding of their metabolism, and particularly of the operation of insecticides, and to study particular infestation problems on an *ad hoc* basis, using radioactive bait to label the insects themselves. This kind of work is often most urgently required in developing areas, where much of it is carried out under the auspices of the International Atomic Energy Agency of the United Nations, or other international aid schemes. The most spectacular success was a tracer investigation into the life-cycle and habits of the screw-worm fly, *Callitraga hominivorax*, leading to its elimination as a major cattle pest in much of the southern United States: the investigation was followed by an all-out attack, in which billions of robust male insects reared in captivity were sterilized by radiation from cobalt-60 and released over the infested area. Such a high proportion of the subsequent matings were infertile that in a few generations the pest was brought under complete control.

Radiation has been used to produce a number of economically important mutants in crop plants. By now, millions of hectares of land carry crop varieties produced in this new way, such as: high-yield wheat; and high-yield rice with good response to fertilizers (a radiation-induced mutant of rice has been produced which contains twice as much protein as the present variety and matures two months earlier). Radiation is potentially valuable (though as yet scarcely exploited) as a means of prolonging the storage life or improving the safety of food and other agricultural products.

Industry

Tracer applications similar in principle to those used in medicine and agriculture are used in industry on many kinds of research problems in physics, chemistry, and metallurgy, and on *ad hoc* practical investigations under operational conditions. The latter cover studies of process-plant kinetics, gas- and fluid-flow measurements, leak location in pipes, plant, or components, lubrication and wear studies, investigation of the movement of dredged spoil in coastal waters, and the behaviour of underground waters in tips, dams, mine-workings, or geological formations. Instruments based on absorption or scattering of radiation are used for automatic process control and product testing, particularly for continuous thickness or density measurement, for elemental analysis, and for monitoring the level of the contents of storage or process vessels or of retail sale packages. Gamma radiography is widely used as an alternative to X-rays for non-destructive testing of welds and castings, especially where portability and independence from electricity supply is important, for example, on pipe-line work; capital costs are also substantially lower. Radiation from powerful cobalt-60 sources is used for stimulating certain chemical reactions of industrial importance, particularly in the field of plastics, synthetic fibres and organic chemicals, but with few exceptions these are as yet of limited importance. By far the most important application of nuclear radiation on an industrial scale is the cold sterilization by gamma rays from cobalt-60 of pre-packed disposable medical goods, especially syringes, needles, blades, sutures, catheters, and a range of pharmaceutical products. This process has revolutionized certain sections of the medical supply industry, and by greatly simplifying clinical procedures (and largely eliminating cross-infection from re-use of instruments) has substantially helped the hospital services and the medical and nursing professions, in the U.K. and in the growing number of European countries that have purchased irradiation plant.

Power sources

Certain radioisotopes, such as tritium (hydrogen-3), strontium-90, and plutonium-238, can provide limited amounts of useful energy over long periods without attention. This property has been exploited for many years in luminous paints and is now used in cinema and aircraft-cabin signs. Power units for light or radio beacons, for telecommunications relays and satellite transmitters, and for implanted cardiac pace-makers, are already in limited use. As energy-conversion techniques become more efficient, and as costs come down with increasing production, these devices appear likely to attain considerable practical importance, particularly in medicine and in the communications field.

The use of nuclear techniques in the development of natural resources

The increasing consumption of raw materials which has characterized industrial development throughout the world during the past century has given rise to a systematic increase in the effort applied to problems of exploration, exploitation, and refining of minerals. In recent years, particularly as a result of intensive developments in nuclear and radiation physics, nuclear geophysical methods of analysis and of investigating mineral deposits have begun to acquire increasing importance and successful applications have now been established in exploration, in controlling production in oil and gas wells, in mine evaluation and development, and in mineral processing. A wide variety of nuclear techniques is already in use for these applications, and a considerable potential exists.

Exploration

Borehole logging is the main application of nuclear techniques in the oil and gas industries. This is a technique whereby instruments are lowered into a borehole so as to obtain direct information on the nature and composition of underground strata. For this purpose, instruments which measure the natural gamma radiation from rocks are most widely used; these give information on rock type and thickness of strata and enable stratigraphical correlations to be made between adjacent boreholes so that particular strata can be followed, sometimes for hundreds of miles. These instruments provide a measure of the potassium, thorium, and uranium contents of the rocks, all of these elements exhibiting natural radioactivity; as the relative proportions of these elements depend on the sedimentation conditions under which the rocks were formed, the ratio of the concentrations measured by the instrument can be characteristic of particular deposits.

Information on the density of strata is required in order to calculate rock porosity, which is of fundamental importance in determining the value of a potential oil-bearing strata. For this purpose a technique known as 'non-selective $(\gamma-\gamma)$ logging' is widely used, sometimes in combination with other techniques. In this technique, a primary gamma-ray source is used with a gamma-ray detector to measure the gamma radiation scattered by the constituents of the rock. Neutron logs, which follow a similar principle with a neutron source instead of a gamma-ray source and neutron or gamma-ray detectors in different variants, are used mainly as an indication of hydrogen content and therefore of the porosity of fluid-filled formations.

Accurate location of the boundary between strata bearing oil and water, gas and water, or gas and oil is a very important problem, both during the evaluation and the exploitation phases of oil production. Borehole probes based on a neutron source, and measuring the intensity of prompt gamma radiation from chlorine present in water but not in oil, were developed for this purpose. However, with the availability of pulsed-neutron generators capable of being inserted in boreholes, the prompt gamma technique has been superseded by the neutron-die-away technique. This relies on measuring the rate of decay of a pulse of neutrons injected into the strata (in about 10 microseconds), the pulse 'dying away' much more rapidly when water is present.

Nuclear techniques have found a particularly important application in the exploration for uranium and thorium, the basic raw materials of the nuclear power industry. Portable field survey equipment, incorporating Geiger-counter detectors and capable of being carried by one person, found early use both by prospectors and by enthusiastic amateurs. However, most prospecting for uranium is now done by mining and exploration companies using the most sophisticated methods. Portable Geiger-counter equipment is still used, but this is supplemented by portable scintillation counters, semi-portable gamma-ray spectrometers, borehole probes, and airborne gamma-ray spectrometers with varying degrees of energy resolution. On the ground, these nuclear techniques are used to detect the presence of radon (the radioactive inert gas formed from uranium by radioactive decay) in soil gas and in surface water and so to establish the presence of uranium, as well as making direct measurement of the radioactivity of rocks. A typical operational sequence for uranium prospecting, once a favourable geological environment has been established, is to carry out airborne surveys and to follow this up with detailed geological mapping and ground surveys using scintillation spectrometers, so as to establish localized regions in which surface trenching or drilling could be worthwhile.

The use of nuclear techniques for metalliferous mineral exploration is not yet established to the same degree as for naturally radioactive materials and for oil, but the possibility of their use is currently attracting a great deal of attention in various parts of the world and several important applications are emerging. For general prospecting, where interest is in elemental concentrations of a few parts per million, various neutron techniques appear promising. Recently, borehole probes using these techniques have been reported from Australia, while in the Philippines neutron activation analysis has been used to establish the occurrence of gold-bearing regions.

Because many instruments incorporating nuclear techniques can be made extremely rugged, and measurements can be made through thick steel walls, these techniques are likely to have important potential application in mineral exploration on the sea bed.

Mine development and control

The most important technique in this area of application is radio-isotope X-ray fluorescence, which is the underlying principle of a number of different types of portable spectrometers now in general use both above and below ground. The limits of detection of radio-isotope spectrometers generally lie between 0·1 and 0·01 per cent elemental concentration and this is consistent with the limits of economic cut-off of many mineral deposits. Of the various applications of these instruments, most are concerned with speeding up or reducing the cost of decision-making processes relating to the directions in which working areas in a mine should be extended, or in deciding whether or not the average grade of an ore is high enough to justify processing. The recent development of radioisotope X-ray fluorescence borehole probes promises to increase further the use of this technique in mine operation. In the coal industry the automatic control of mechanical miners operating on a 'long wall' coal face has now been achieved by using a gamma-backscatter technique with radioisotope sources and scintillation counters mounted above or below the machine and sensitive to the presence of coal or rock. With these devices, which are now coming into general use, undulations in a coal seam can be followed automatically, so that the probability of extracting rock from the floor or of removing the layer of coal supporting the roof is greatly reduced.

Oil extraction

In many cases the recovery of oil is accompanied by an upsurge of water which enters the oil-bearing formation from an underlying aquifer as a result of a reduction in pressure as oil is withdrawn from the well. If such a possibility exists, it is extremely important that the movement in the oil–water interface can be established and the rate at which oil is removed can be controlled so that water is never pumped out of the well. The neutron-die-away technique, referred to above, has found important use in this application.

A variety of techniques using radioisotopes as tracers are also in routine use, but probably the most important is in the secondary recovery of oil, when water is injected into adjacent boreholes in order to enhance oil extraction. Suitable radioisotopes are added to the water to give information on the manner in which the water leaves the injection wells.

Mineral processing

Large variations in the elemental concentration of slurries in mineral processing plants may occur due to changes in composition of the raw material at the input to the plant. Efficient control can only be achieved through continuous analysis of the process stream. In general the concentration of one or two elements is required to a precision of between 5 and 10 per cent of the elemental concentration in the slurry solids and this concentration may have values from about 75 per cent by weight in concentrates to less than 0·1 per cent in residues.

Several nuclear techniques are in routine use and several others are currently the subject of active development. The most important of these techniques is radioisotope X-ray fluorescence which is now being used for the analysis of sulphur, silicon, calcium, zinc, copper, molybdenum, tin, and barium ores. Concentrations which can be measured vary from between approximately 0·01 per cent for molybdenum in tailings to 55 per cent for zinc in concentrates and 88 per cent for calcium in cement.

Neutron activation analysis has considerable potential for on-line measurement as the method has important advantages over conventional analytical methods. Mainly, its advantages arise from the relatively high degree of penetration of neutrons and of the resultant gamma radiation, which allows measurements to be made through the walls of pipes, bunkers, and process vessels; on conveyor belts without the need for sample preparation; and, generally, in hostile environments. Analyses tend to be restricted to elements having favourable nuclear properties and these include fluorine, carbon, oxygen, nitrogen, silicon, aluminium, phosphorus, magnesium, copper, iron, and barium. Most of these elements cannot be measured on-line using X-ray fluorescence techniques except with considerable difficulty, so that neutron activation analysis can be regarded as a complementing technique to X-ray fluorescence. Radioisotope X-ray fluorescence has an important application in the measurement of ash in coal, and instruments based on this technique are now in routine use in the United Kingdom and Europe to control coal blending operations.

It will have been realized from this brief and necessarily selective survey that, while atom bombs and nuclear power stations have been the most dramatic manifestation of applied nuclear energy following its discovery in the late 1930s, nuclear techniques have an immensely various and all-pervading potential, and indeed there is every reason to suppose that they will in time cease to be regarded as something 'special' and will merge into the general mass of normal scientific techniques, to the body of which they will add considerably both in numbers and in extent.

The moral aspects

By the Rt. Revd. Robert C. Mortimer, Bishop of Exeter
The concept of morality presupposes the concept of freedom to choose. 'I ought' involves 'I can'. Man has freedom to choose only when he has power. The areas of conduct in which moral judgements are relevant or meaningful are relative to the areas in which man has the power of choice. In those areas he has moral responsibility, and is subject to praise or condemnation, dependent on whether he has chosen rightly or wrongly. Every extension of power, therefore, extends the area of moral responsibility.

New scientific discoveries and advances in technology increase human power and, therefore, moral responsibility. Matters are brought within the range of human decisions which formerly were outside that range. The ability to employ nuclear energy is an immense advance in human power to control and adapt man's environment and an immense increase, therefore, in human responsibility. All power can be used either for good or for evil. The discovery of the ability to employ nuclear energy differs in degree rather than in kind from previous situations. The new power is so vast, and the consequences of its use or misuse are so far-reaching, that we tend at first to be appalled by the tremendous responsibility which it imposes on the human race, and to wish that the discovery had never been made.

But this is a wrong and cowardly reaction. It is the Christian conviction that man is intended to be 'the Lord of Creation', that he is meant to discover 'the secrets of nature' and to apply his knowledge and power to the betterment of his environment, to the improvement or amelioration of the conditions under which human life is lived. To shrink from this, because the power to improve involves also and necessarily the power to worsen, because the power to create carries with it the power to destroy, would be to betray the human race. The right choice is to welcome the new power and to avoid its misuse.

In the case of nuclear energy, the fact that it was first used to destroy and not to create—indeed that the motive in learning how to harness nuclear energy was to possess an irresistible weapon of destruction—has confused us. The destructive power of nuclear energy is only too apparent and its constructive possibilities are less clearly seen. It is, therefore, a not surprising reaction that we should wish that the discovery had never been made, that we should attempt the impossible task of halting the use of nuclear energy in the position in which it now stands and prevent any further advance, and that we should pretend that, by somehow securing the destruction of existing nuclear weapons, we can ensure that the new knowledge will be laid aside and forgotten, or, at least, never employed. We are tempted to restrict any new research into the uses of nuclear energy because of our fear that any further knowledge and power will only increase the possibilities of disastrous misuse.

But there is no simple way of escape in this direction. 'The bomb' exists and the human race has to learn how to live with it. The human race must, from now on, live under the continuing threat or at least possibility of self-destruction. The problem is how to control that threat. The threat derives from the ingrained human tendency to find the ultimate solution of disputes and disagreements in the arbitrament of force. If the worst comes to the worst and no other solution can be found, war is the final resort and victory or defeat the conclusive answer. In the course of human history, resort to war has often been far too easy and frequent. Sometimes it has been the act of a flagrantly immoral and aggressive individual or popular will, with no attempt made to justify it. More often it has been recognized to need justification, but the justification has been no more than a cloak to conceal pride or greed. Sometimes war has been deliberately and reluctantly chosen as the lesser of two evils. Yet because of the essentially evil nature of war—for at best it is only the lesser evil of two—attempts have always been made to limit its savagery and violence by so-called 'laws of war'. The ban on poisoning wells and the sanctity of envoys are examples from earlier human history: the treatment of prisoners, the immunity of civilians and the protection of the wounded are later developments. However much these 'laws' have been broken or ignored, their existence is evidence of the recognition by the human race of the essentially evil character of war.

The existence of 'the bomb' forces upon us a new appraisal of the part, if any, which war may be allowed to play in the solution of human quarrels. The question of how to eliminate war altogether becomes an urgent one. International agreements to renounce aggression and world pacts not to resort to war are no doubt of value, but it would be foolish, as experience has shown, to rely much upon them. To eliminate war altogether it is necessary to eliminate the causes of war. And these lie deep in human nature. They are the greed, fear, suspicion, envy, and pride which are ingrained in every man. It is the presence of these qualities in every man which makes it necessary for all human communities to have their police forces and their penal systems if their members are to live peaceably together. And even so, it is only where in the majority of the members of a society these qualities are under voluntary control and are not allowed to find expression in violence and law-breaking that a peaceable coherent community is possible. Yet within all communities crime and violence exist, because the evil impulses from which they spring are under imperfect control, and if the control breaks down not merely in the case of an individual here or an individual there, but within a large section of the community, anarchy and destruction follow. It is the same as between nations. The pride, anger, suspicion, envy, and greed which exist to a greater or less degree in every human individual being, can, under provocation, take possession of whole communities. These can develop a communal hatred and suspicion of one nation for another, a common determination on the part of one nation to force its supposedly just and righteous will on another, and on the part of that other a common determination at almost any cost in violence to resist any such attempt. Thus a war hysteria develops.

To change this situation, to eliminate the deep causes of war, requires more than good resolutions made in times of peace and freedom from stress. It requires a deep change in human nature. It requires what religion calls repentance, a complete change of heart, of approach to life on the part of all or at least the vast majority of human beings. A change from that self-regard, which is the essential food from which all our greed, fear, and hate draws its strength, to a regard for others, to a natural and spontaneous consideration for the needs and claims of others, and an acceptance of the cost to ourselves of meeting those needs and claims.

Such a change of heart and outlook on the part of the human race as a whole seems impossibly remote. Indeed the tendency would seem to be all the other way. The preoccupation of the developed countries with affluence, the maintenance of standards of living and individual status, reinforces the natural human instincts of greed and self-regard. Yet above all else the human race needs to return to a deeper acceptance of the teachings of the ancient religions and philosophies, namely, the dependence of man upon a Being and a Power outside and other than himself, which calls for obedience, humility, and a certain detachment from purely material considerations. What is required is a general agreement that in international affairs there are shared human values linked to inalienable natural rights which cannot be alienated by any human authority, but which are dependant on the Will of the Creator expressed in laws recognized by the human conscience. And this recovered awareness of a 'natural law' must be accepted as the basis on which all reasonable differences and disputes are to be discussed and settled in a reasonable manner, peaceably and without harm to mankind. The attitude of the human race, trying to learn to live with and overcome the threat of nuclear destruction, should be that of the humble despairingly hopeful cry 'Lord, if Thou wilt Thou canst make me clean'; to which surely, if the cry is sincere, will come the answer 'I will, be thou clean'.

Meanwhile the threat has so far been contained by other means. Restraint has been imposed, not by the release of new powers of altruism in the human race, but by reliance on precisely those elements of self-regard and fear which give rise to the threat in the first place. If the good in man is too weak to enable him to find the means of self-preservation, the bad in man shall be called in to his rescue. The great powers are restrained from nuclear war with each other by the mutual fear that each can only succeed in destroying the other. Possession of 'the bomb' has come to mean only the possession of the means of mutual destruction. The impulses of hatred, anger, and greed are halted by fear. This fear exercises its restraining force not only where two opposing states each possess the power of destroying the other, but also where only one of them does. The possessing power dare not use its nuclear weapons, partly for good humanitarian reasons, partly from fear of world public opinion, which would undoubtedly be outraged by such massive destruction, and partly by fear of the unpredictable dangers to itself of the consequent atmospheric pollution. On the other hand, the power which does not possess nuclear weapons is restrained from pursuing its war aims to the point of 'unconditional surrender' or even to the point of seriously damaging any one of its opponent's vital interests by the fear, lest by so doing it should provoke the use of that opponent's nuclear weapons.

Thus the existence of 'the bomb' has had a restraining effect. Its possession has prevented war or has limited and localized wars. And yet 'the bomb' itself is utterly evil, and its use in any circumstances would be morally indefensible. And this for three reasons. First its completely indiscriminate nature and the widespread havoc its use would entail put it beyond the pale. Second, the consequent atmospheric pollution would have serious consequences not only, even if chiefly, for the nation at which it was aimed, but also for neutral nations, indeed for the whole human race. And third, and most important of all, the precisely unpredictable but certain consequences it would have for generations yet unborn. The chief difference between a nuclear and non-nuclear war is this. A non-nuclear war ends with the ending of military operations. The work of reconstruction can begin, and in time the destructive consequences of the war are obliterated. But the consequences of a nuclear war do not cease with its ending. The atmosphere remains polluted, and its harmful effects upon the human gene-pool cannot be halted or undone. Not this generation only, but succeeding generations suffer. And these considerations apply not only to the use of 'strategic' nuclear weapons, but also to the so-called tactical ones. Quantitatively they may be distinguished, but qualitatively they are the same. It is to be hoped that nations will not allow themselves to be duped into believing that tactical nuclear weapons can be justified like the illegitimate baby, by the plea that 'it is only a little one'. But it is, perhaps, unlikely that tactical weapons will be used, for fear of more massive retaliation and ultimate inevitable escalation into total nuclear war.

This reliance on fear to contain the threat of nuclear destruction is negative and risky. A madness of self-immolation might at any time take possession of a people or its political rulers. There is an urgent need to discover and implement some international machinery which could make it practically impossible for any one nation or its leaders, by unleashing its nuclear destructive powers in a moment of passion, to risk the destruction of the human race. But the nations, engrossed as they are by the importance of preserving their national sovereignties in fact seem strangely blind to their danger and their need. The first duty of national statesmen today is to concentrate their energies on the devising and creation of some effective international instrument of restraint on the absolute autonomy of national sovereign states.

The possibilities for evil which result from the discovery of how to employ nuclear energies are clear enough, but they must not be allowed to obscure the immense possibilities for good which also result. These possibilities are described elsewhere, and their development in the interests of the entire human race is a clear moral duty. But it is development in the interests of mankind in general, not of particular individuals or nations. It would be only too easy, since such development presupposes great economic resources, for the end result to be yet a further increase in the gap between the advanced and the backward peoples. But the vast new powers now coming within man's grasp can and should be used for the benefit of all. Indeed, since the development of these powers must inevitably bring human societies into a closer relationship with one another and cause a greater realization of their mutual interdependence, economic and political competition will add incentives towards a greater solidarity between rich and poor nations. As all become more and more mutually interdependent, the interest of one becomes the interest of all. The employment of nuclear energy to enrich the poorer nations and ameliorate the conditions of life for their peoples is in the long-term interest of the richer nations as well as being their duty.

Inevitably the employment of nuclear power to release the natural resources and riches of the under-developed parts of the world will bring about radical changes in the social and political lives of the people who inhabit those parts. They will be transformed into educated, technological, sophisticated, and affluent societies. Though the ultimate responsibility for the quality of life which will result must lie with the peoples themselves, yet there is a moral duty on the rich nations to do what can be done to help the developing nations to avoid the mistakes and faults which are apparent in their own older societies and an alertness to perceive what lessons they themselves can learn from the newly developing ones. There are or may be human qualities in primitive societies which can and should be preserved as they develop, and which could be grafted into the life of the more advanced societies to their own immense advantage.

The development of nuclear energies for peaceful purposes is not without its risks. Radioactive materials are extremely dangerous. They involve serious risk both to those who actually handle them and to all those who come to be exposed to their contamination. Radiation protection is therefore a matter of the greatest importance, and it is a paramount moral duty to do everything possible to ensure that the risk of exposure is made as low as possible. But some risk there must always be. There is risk in everything. The problem is to discover what is the acceptable level of risk. This level is always relative to the benefit conferred by that which gives rise to the risk: the greater the benefit, the higher the level of risk which is acceptable. The use of coal and oil for the purpose of generating power is an ever-increasing public hazard. Yet, until some alternative source of power is discovered, a high level of risk is preferred to the abandonment or even drastic reduction of the employment of coal and oil.

The measures needed to give adequate protection from radiation exposure to the workers in industries creating or using nuclear energy, and the level of risk to which they may be allowed to be exposed, involve highly technical considerations which it is beyond my competence to discuss. As a general principle it may, however, be safely asserted that radiation work should not be more hazardous than other 'safe' occupations, and in consequence that those exposures which would certainly cause harmful effects should be prohibited. In fact, there are some internationally agreed limitations of occupational exposure which confine the individual risk to what is regarded as an acceptable limit. It is a more difficult problem to reach agreement on what should be the limit of risk for people not occupationally exposed; for example, if further nuclear explosions are necessary, if the means of fully exploiting the peaceful uses of nuclear energy are to be found, what level of release of radioactive substances into the atmosphere would be regarded as acceptable?

The risks inherent in the development and use of nuclear energy are not only personal, but social; and not only social but genetic. The decision to incur these risks cannot, therefore, be left to individuals or private industrial concerns. The welfare of mankind demands that there be an international authority which shall lay down the appropriate conditions and limitations for work and exposure, both for operatives and for the general public. It should be incumbent on all national authorities to see that these conditions are strictly observed.

If there is, as I believe, a moral duty to develop the peaceful use of nuclear energy, there is also an equal duty to see that the risks involved are reduced to a minimum. The possibilities both for good and for evil in the use of nuclear energy demand an effective international authority. For the good purposes the establishment of such an authority should not be all that difficult, nor obedience to it reluctant. For the evil purposes, the creation of such an authority is infinitely more difficult, and obedience to it harder to obtain. Its necessity and urgency, however, is even greater.

6

6

The sociological consequences of nuclear energy

By Lord Ritchie-Calder

Popular misgivings about nuclear energy, and the social and economic restraints which they impose, belong less to sociology than to mythology.

In October 1957 the World Health Organization set up a study group on the mental health aspects of the peaceful uses of nuclear energy meeting at Geneva. The group, which was composed of physicians, physiologists, psychiatrists, psychologists, behaviourists, and sociologists, had reports before it from investigators in many parts of the world, not only in sophisticated societies but in the less developed countries. Although its terms of reference included the brain, in the physiological sense, the measurable effects of actual or potential damage were so insignificant, quantitatively, as to be disregarded: 'mental' became 'emotional'. The evidence showed universal disquiet about nuclear energy, not only about its cataclysmic possibilities in a nuclear war, but about its peaceful applications. It appeared to the W.H.O study group that the crust of our civilization is only eggshell thick. The group agreed that, confronted with immeasurable power from the infinitesimally small nucleus, civilized man tends to cower, like his Neanderthal ancestors, 'in the dark caves of his own emotions' or, in the less poetic terms of the psychologists, 'The tendency to relapse into more primitive forms of thought and feeling which is characteristic of most of the reported reactions of the public to nuclear energy can be ascribed to a psychological mechanism known as "regression".' Our primitive ancestors dreaded, and sought to appease, the elemental gods of thunder and lightning and fire, but the elemental gods of the new mythology are radioactive, unseen, unsmelt, unfelt, unheard, and all-pervasive. They are also man-made. In our retreat into 'the childhood of mankind' (to quote the study group) we revive a cosmic guilt about tampering with things we should not. This exists in the myth and legend of every culture: Adam and Eve eating the forbidden fruit of the Tree of Knowledge; Prometheus stealing fire, the prerogative of the gods; Pandora opening the box; Faust trafficking with the Devil; and the Ancient Egyptian saying, 'When Man learns what moves the stars, the Sphinx will laugh and all life upon earth will be destroyed.'

All this is quite irrational, but if we consider the reactions and reflexes of educated people, including guilt-ridden eminent scientists, of judges ruling on the siting of reactors (e.g. the Fermi reactor on Lake Michigan), of reasonably well-informed journalists, or of spokesmen at public inquiries, this new 'mythology' is not to be discounted. Even the nuts-and-bolts, or 'nitty-gritty', men who have to plan and build new installations cannot ignore this mythology, although they may call it 'ignorance' or 'stupidity'; it is the imponderable in their feasibility studies.

The misgivings can be explained without going back to 'the childhood of mankind'. The first the world knew about the release of nuclear energy was when the bombs destroyed Hiroshima and Nagasaki. The whole project had been contrived behind impenetrable secrecy fences, to culminate in an apocalyptic explosion. Some of us may have known or sensed the significance of Hahn and Strassmann's recognition of uranium fission and Meitner and Frisch's prognosis of chain-reaction. Some might have followed the flurry of scientific papers in 1939 and 1940 in which discoveries and confirmations (and warnings) followed each other in quick succession until security (in the first instance self-imposed by the scientists) closed down the bazaar-and-mart of interdisciplinary scientific exchanges. From there on, and further reinforced by the military security of the Manhattan Project, nuclear energy became the 'physicists' bomb'. Security classification was confined to the physicists, chemists, and engineers engaged on the project. Scientists in other disciplines were not consulted and certainly not the sociologists.

Clement Attlee, who, as British Prime Minister had concurred with President Truman in the decision to drop the bomb on Hiroshima said later:†

> We knew nothing whatever at that time about the genetic effects of an atomic explosion. I knew nothing about fall-out and all the rest of what emerged after Hiroshima. As far as I know, President Truman and Winston

† *A Prime Minister remembers* (ed. F. Williams), Heinemann, London (1961).

Churchill knew nothing of these things either, nor did Sir John Anderson, who co-ordinated research on our side. Whether the scientists directly concerned knew, or guessed, I do not know. But if they did, then, as far as I am aware, they said nothing of it to those who had to make the decisions.

A strange lapse! Hermann J. Müller had shown in 1927 that chromosomes could be altered by X-rays, and induced genetic mutations were familiar to the biologist. If the men of affairs were uninformed about possible hereditary effects, the world at large was completely ignorant until the two cities were destroyed. People died from the effects of radiation and to it all was added the awesome realization that the effects could be visited upon future generations. It was this that added to atom-bombing a dimension of horror greater even than the massive high-explosive and fire raids of World War II.

Thus the release of nuclear energy, Man's greatest material achievement since the mastery of fire by our remote ancestors, burst upon an uninformed and uncomprehending world and all the attempts to explain, let alone justify, only aggravated the suspicions and exaggerated the distrust. Nuclear superstition, as the W.H.O study group found, pervaded the world, in advanced and primitive communities alike. The only way to rebut superstition is to confront it with reason, but what happens when the repositories of reason are not believed?

The University of Chicago initiated in 1957 a series of inquiries into the social implications of nuclear energy, which continued for nearly ten years. From the outset those concerned realized that scientific veracity had been impugned by circumstance. The scientist had been responsible for the achievement which had induced the fears and he was being asked to explain away those fears. He was being called upon not only as the expert witness but as the spokesman, not only to give the facts, within the acknowledged limitation of his specialized knowledge, but to extrapolate these facts beyond his knowledge and to pass judgements and express opinions. One of the difficulties is that to the public 'scientist' is a generic term; the ordinary person does not distinguish between a physicist, a chemist, or a biologist. A physicist, dealing with exact measurements, sounds categorical. A biologist, dealing with many more variables, can be frank about his reservations. Each may be stating facts with scientific honesty but, in the emotion-charged area of atomic energy, when a biologist queries or qualifies the position taken by the physicist, the ordinary person concludes that 'the scientist has contradicted himself'.

The fact that scientists, and the authority of science, were used to promote policies, to win appropriations, to defend government agencies and industrial concerns, and to 'reassure' the public on subjects such as fall-out during the atmospheric bomb-testings tended to make people suspicious of the scientists' motives and of the validity of their facts. When world-renowned scientists ranged themselves in the public debate it was a toss-up which authority to believe. Furthermore, scientists experimenting with the biosphere, our living-space, could be wrong for all the world to see, or they could be caught suppressing facts. There was the notorious case of the *Fukuryu Maru* (*Lucky Dragon*). The bomb tests at Bikini Atoll on 1 March 1954 resulted in fall-out of radioactive dust on a Japanese fishing-boat well outside the test area within which (it had been confidently stated) any hazards would be circumscribed. Apart from the radioactive burns to twenty-three fishermen, one of whom died, the fish hauls in the South Pacific were radioactively contaminated. (The U.S. Government paid two million dollars in compensation.) High radioactivity was found in tuna and other fish brought into Japanese harbours. No arguments in mitigation could alter the fact that 'the scientists miscalculated'.

When the H-bomb was exploded we were assured by the responsible scientists that the fall-out-fission from the fissile detonator would be 'purely local'; the thermonuclear explosion would punch a hole in the stratosphere and the radioactive gases would be dissipated above the tropopause. That left two factors out of account: first, that radioactive krypton decays quickly into radionuclides of strontium (^{90}Sr) with a half-life of 27·7 years; second, that the tropopause is not continuous. Through the 'fanlight' of the tropopause, the radio-strontium returned to our atmosphere, to be spread by the climatic jet-streams all over the world and deposited in rain.

One recalls the hullaballoo about radiostrontium when people suddenly became aware of it as the bone-seeking radioisotope, becoming incorporated into the skeletons of growing children. Nor did it help when it was discovered that the facts and effects of radio-strontium had been known for a long time but had been classified as secret. Professor Cyril Comar had identified it at Oak Ridge in 1948 in the burns of the hides of animals from the Nevada proving grounds. He had recognized it as a bone-seeker and had reported his findings. Because they would have caused alarm and despondency, they were classified as secret and his researches were put under wraps as

'Operation Sunshine'. When the radioactive chickens came home to roost, the facts could no longer be concealed.

The military and peaceful uses of nuclear energy thus interacted in creating an atmosphere of alarm and mistrust and, on the authorities, side, excessive secrecy. (What they could not explain, they classified as secret.) This was not conducive to the kind of dialogue which was necessary if full advantage were to be taken of the social and industrial possibilities of a new source of energy.

On the other hand, 'nuclear superstition' probably had one permanently good effect: Hiroshima and the fall-out registered radiation as something not to be trifled with, and when it is boasted (and with justification) that the nuclear industry is the safest of all industries, in terms of health and accidents, this is due to the fact that the scare-factor was built in the precautions from the outset. Since the concern was universal, the imposition of international standards was accepted. When, in the light of years of experience, some of the precautions are claimed to be excessive, their modification is about as difficult as amending the American constitution. This is, socially, a good thing. The safeguards have been embodied in the economics of nuclear energy. If this had happened in traditional industries a great deal of what we call pollution and human sacrifice through fatal accidents, permanent injury, and ill-health could have been avoided. And since the safeguards are internationally accepted the excuse 'Our competitors do not have those restrictions' does not apply.

The first substantial breakdown of barriers was the United Nations Conference on the Peaceful Uses of Atomic Energy in 1955. For one thing, it showed that a great deal of the secrecy was absurd because, without access to the 'secrets' of the nuclear powers or, indeed, without espionage, scientists all over the world had arrived at similar findings. Scientists who had been isolated by national security systems were again on speaking terms with their colleagues and could discuss systems by which nuclear energy could be industrialized. In this state of nuclear euphoria, the threat became a promise. Instead of mankind going to be extinguished by the Bomb, it was going to be redeemed by nuclear power. Nuclear energy, unhampered by geology or geography, would bring industrialization to the developing countries. The handicaps of lack of fossil fuels or water-power would be overcome and those who had not had the advantages of the Industrial Revolution would 'make a leap across the centuries'. Inaccessible regions of the world would be opened up to industrialization. The jungles, the deserts, and the frozen wastes could have packaged reactors to exploit any natural resources which might exist, without waiting for the construction of railways or roads as had been the historical pattern. The components would be flown in by transport planes (as had been done to service the military operations during World War II). Once the reactor was assembled, fuel replenishment would be simple —there would be no need for continuous supplies of coal or oil by freight trains, trucks, or pipe-lines, simply periodic substitution of enriched fissile fuels in comparatively small bulk. We talked about 'industrial oases' in, for example, the Arctic regions, where mineral wealth could be extracted and refined on the spot and flown out in concentrated forms from all-the-year round air strips, kept frozen, like ice-rinks, during the freeze-up and the break-up, with the help of surplus nuclear power. Whole areas would thus become part of our habitable world not by massive civil engineering but by organic growth from the inside outwards.

At the first U.N. Atomic Energy Conference in 1955 (but less so at the 1958 Conference) one would have imagined from the breezy salesmanship of such high hopes that if a developing country wanted a packaged reactor it was just a case of having it wrapped up to take away. Even more exciting possibilities were cited. The President of the 1955 Atomic Energy Conference, the late Professor Homi J. Bhabha, first publicly referred to the use of thermonuclear energy for civil purposes ('putting the H-bomb into dungarees' was one journalistic phrase) and he promised 'as much power as there is deuterium in the seven seas'. Provoked by this statement, the British, Americans, and Russians all admitted, prematurely and over-optimistically, that they were working on to thermonuclear energy. Here we were being offered not only the 'taming of the H-bomb' but 'clean' energy, without the problems of disposal of atomic waste from fission reactors. As an antidote to nuclear superstition this was strong medicine: fusion energy was the building-up of atoms as contrasted with fission energy, the breaking-down of atoms. Psychologically, this was construction, not destruction. Those were the halcyon days: nuclear camaraderie, fences down, high promises.

The general impression, encouraged further by the setting-up of the International Atomic Energy Agency, was that the under-powered countries were to make their 'leap across the centuries' with the help of nuclear reactors 'tailored' to their needs. With small or moderate-sized reactors, their industries would be able to grow (and the reactors would grow with the industries, rather like the factory power-house giving way to the local generating station and finally to the Grid). How this was to be done before the countries had trained nuclear scientists and nuclear engineers was conveniently overlooked. It has not in the event been achieved, except in a few countries which are an embarrassment in the discussions of nonproliferation.

No one questioned the wisdom of necessity which dictated that Britain should supplement its energy supplies from conventional stations by installing natural uranium reactors of the Calder Hall 'Magnox' type. This was the starting-point of a programme which by 1970 was producing 5000 megawatts of electrical capacity. The Second Nuclear Power Programme announced in 1965 was for 8000 megawatts with the emphasis on the advanced gas-cooled reactor, fuelled with ceramic uranium dioxide, which allows higher fuel temperatures and higher fuel ratings, with greater irradiation and compactness. These are again of the order of 600 megawatts for each twin-reactor. This is splendid for the needs and economy of the Central Electricity Generating Board system, the biggest single network in the world, but even a single 600-MW reactor has an output too large for many other countries. A 360-MW station has therefore been designed for export incorporating an AGR reactor in a system based on turbines and generators and other components familiar in conventional power stations. Standard parts and simplified construction enable local industry to make a substantial contribution to the assembly. This is an eminently sensible approach but one must be grudging and suggest that it is still, in terms of the needs of developing countries, a case of 'unto him that hath shall be given'. A locality capable of absorbing a base-load of 360 MW must already have a significant industry whereas a locality *needing* industry would want something smaller. (Notice that one is talking here in sociological and not in economic terms. Reactors big or small have somehow to be paid for.) There would thus still seem to be an international relations argument for packaged reactors of much smaller capacity.

When we talk about fast breeders, of course, we are in the realm of what still to most people is fantasy, however real and practical the system is to the scientists and the engineers. The 'pixilated pile', which produces fuel as it consumes it, is like the fairies filling the coal-scuttle every time it is emptied. But it is a fantasy without nightmares.

The record of the 'atoms industry' in the quarter of a century which has followed the release of nuclear energy has been reassuring. No accident which has occurred has involved great loss of life or serious damage. From the point of view of industrial health and accident rates it is the exception and a model for every other industry. The fact that people still say 'But what about Windscale?' illustrates this exemplary record. As a 'disaster' it was a non-event. It nevertheless showed the strength of 'nuclear superstition' and how difficult it is to counteract it. In October 1957, a reactor 'burned out' at the Windscale plant of the U.K. Atomic Energy Authority. The reactions of the world press were illuminating. The fire had been dealt with by firemen—daring firemen, but nevertheless firemen, a familiar concept to the public. On the first day, the incident was given a lot of newspaper coverage— big headlines and extensive detail—but even the more sensational newspapers were factual and reassuring in their presentation. Here was the proof that a reactor did not explode like a bomb and that firemen had coped. On the second day, the Atomic Energy Authority, very properly, had sent out monitoring squads to sample the herbage, the water, etc., in the surrounding countryside. This again made big stories but they were still not alarmist; they gave the facts and treated the monitoring as a sensible precaution by the Atomic Energy Authority. On the third day, the headlines exploded: the Atomic Energy Authority had collected hundreds of gallons of milk because traces of radioactive iodine had been found in local supplies. The milk was dumped in the sea. The key word was 'milk', as it was repeatedly in the alarms about the fall-out from bomb-testing. It was, according to the W.H.O. Study Groups on the Mental Health Aspects of Atomic Energy, a classic example of psychological regression (or, in the phrase of the Chairman, Professor Hans Hoff, the psychiatrist, 'the editors must have been breast-fed'). The A.E.A. had from the outset kept the press well informed and the press, as far as the purveying of facts went, had acted sensibly, but the general effect had been to spread alarm and evoke regressive instincts out of all proportion to the measurable risks. This was not a case of misinformation; it was a case of information given to a scientifically ill-prepared public, which with its instinctive fear aroused would have distrusted official reassurances of any kind.

This crisis of confidence continues in Britain and the United States and (deriving as it does from 'the childhood of mankind') probably in the U.S.S.R. as well. It makes the nuclear experts too categorical or encourages them to blind the public with science. The case of 'm.p.d.'—maximum permissible dose—is an example of how not to 'comfort' the public. Lord Cherwell's public utterance during the alarm about radiostrontium in fall-out was typical: in terms of m.p.d. of radiostrontium, he said, the amount in fall-out was less dangerous than walking 200 feet up a hill, i.e. relative to natural cosmic rays. This professional wisecrack misfired badly on two counts: those who did not know much about m.p.d. resented it as deriding their genuine concern, and those who did know about m.p.d. resented its misuse. Permissible doses belong not to science, but to the 'philosophy of risk'. They have no more scientific authority than the 40 miles per hour speed limit, which has no relation to the capacity of a car but only to a social convention which decrees that it is unwise (and punishable) to travel at more than that speed. In terms of those vocationally engaged in radiation activities, 'm.p.d.' is entirely applicable and highly commendable, but in terms of the population at large it is entirely and dangerously misleading. As international discussants eventually agreed, no radiation dose in excess of natural background radiation is 'permissible'. In the event of a 'blow-out' like the Windscale incident, the idea that there should be no need to worry until the 'dose' exceeds so much would be reprehensible; every corrective measure should be employed to secure a return to zero.

In our concern nowadays about environmental pollution, about smoke and fumes and effluent, nuclear energy is recommendable as a 'clean' and 'safe' source of electricity. Relatively speaking, it is. Its advocates become fretful about public mistrust and impatient about opposition to the siting of generating stations and the disposal of tolerable amounts of low-level wastes in open waters, or about public anxiety over thermal pollution from nuclear installations. After a quarter of a century of reliable performance, they are probably justified in claiming that the precautions are excessive and economically burdensome and could safely be modified. They ought, however, to be grateful for social censorship and for the constraints, because as the uses of fission energy increase, so do the accumulative risks. It is best to think and think again. As in road-safety, you may be a careful driver, but what about the other fellow?

Apart from the operational risks against which such elaborate precautions are taken in design and construction and day-by-day drill, the biggest problem is the disposal of radioactive wastes. Until thermonuclear power is available (and it will bring its own problems of hazards of a different kind and order) the fission products will have to be disposed of. Marine disposal of radioactive 'trash' is being seriously challenged by a new generation of 'environmentalists'. The burial of highly radioactive wastes on land has already reached massive proportions. It can be said that in a quarter of a century it has cost more to bury live atoms than it cost to bury, over 300 years, the pyramid kings of Egypt.

In 1968 the U.S. Atomic Energy Commission asked for $2 500 000 to replace failed and failing tanks at Richland, in the State of Washington and it was stated: 'A total of 149 tanks with about 95 million gallons capacity have been built. Prior to 1964 five tanks had been withdrawn from service because of leaks. Since 1964 one tank has failed and four more have developed indications of incipient failure. The waste storage situation has been further aggravated by recent temperature control problems which have occurred in a tank filled with PUREX self-boiling wastes. The tank bottom temperatures began exceeding the design control limits and further filling of this tank has been stopped which reduces the planned waste storage space.'

Many kinds of disposal have been devised and practised. The wastes can be calcined; they can be vitrified by being converted into glass; they can be stored in limestone caves in disused salt mines or, more unreliably, in exhausted oil-wells. Another method is hydraulic fracture. This involves methods familiar to petroleum engineers and consists of injecting liquids, under pressure, horizontally into shale beds so that the shale splits like cleaving a lump of coal. The liquid in this case can be cement containing radioactive materials. When the cement sets, sandwiched in the shale, it is secure even in earthquake conditions, since it will be displaced without leakage or seepage.

Occasionally the public has a startling reminder of the traffic in radioactive wastes, as in the story of Palomares. In that event, a nuclear bomber collided over the coast of Spain with the flying tanker which was refuelling it and four bombs got loose without, fortunately, a catastrophic explosion. One shattered bomb, however, contaminated the beaches and farm land, and the sand and soil had to be collected into 4810 steel drums and shipped for burial under 10 feet of earth at Savannah, Georgia.

In the United States, the carrying of radioactive wastes from atomic plants to approved burial grounds entails trips amounting to over two million road-miles a year. By the year 2000, if present practices continue and planned installations are developed, the number of six-ton carriers in transit at any given time will be 3250 and the amount of fission products on the highways will be 980 million curies. That is a vast amount of radioactive material to be travelling about a populated country.

The public has got to get used to living with the atom, but for everybody's peace of mind, we need to restore confidence between the people and the expert, even if we have to condone nuclear superstition.

Index

Aberystwyth, University, 53
accidents, nuclear, 4, 35, 36, 60, 79–80, 82–3, 84, 85; *see also* hazards, nuclear; safety regulations
agriculture and nuclear power, 5, 57, 58, 59, 60, 62–3
Alamogordo, 24
alpha radiation, 8, 9, 15
aluminium smelting, 59
ANDERSON, Sir JOHN, 78
Anglo-American co-operation, 1, 25, 27, 28–9
Arabian Gulf, 57
Arctic regions, 81
armaments, *see* weapons
atom first split, 9, 10–11
Atomic Energy Agency, United Nations, 62, 81
Atomic Energy Authority, United Kingdom, 1, 83
Atomic Energy Commission, United States of America, 85
Atomic Energy, Joint Committee of United States Congress, 51
United Nations Conference on Peaceful Uses, 80, 81
World Health Organization Study Group on Mental Health Aspects, 83
atomic mass scale, 8, 10, 11–12, 17, 18–19, 23
atomic number, definition, 10
atomic structure, 7–13
'Atoms for Peace' conference, 3
ATLEE, CLEMENT, 77

Barn, 13
BECQUEREL, HENRI, 7
BEPO (British experimental pile), 16
beta radiation, 9, 14, 18, 19, 20

BHABHA, HOMI J., 81
Bikini Atoll tests, 79
blast, 55–6
BOHR, NIELS, 17
borehole logging, 64–6, 67

calcination of nuclear wastes, 85
Calder Hall nuclear power station, 82
Canada, 25, 37–8, 39
cancer therapy, 22, 61–2
carbon, as moderator, 15, 16, 21; *see also* graphite
energy release, 8–9
carbon-14 in agricultural research, 62
cardiology, 5, 64
cattle pest control, 62–3
Cavendish Laboratory, Cambridge, 10
centrifuge method of uranium enrichment, 29, 44
CHADWICK, JAMES, 9, 10, 26
chain reaction, 14, 15–17, 25–6, 27, 28, 31, 54
CHERWELL, Lord, 84
CHESHIRE, Group Captain LEONARD, 1–2
Chicago, University, 78
CHURCHILL, Sir WINSTON, 77–8
civil engineering, 55, 56, 63
coal, costs, 3, 44, 45, 47, 48, 49, 51–3
energy yield compared with uranium, 14
hazards, 75
nuclear techniques in mining, 67, 68
transport, 37
waste disposal, 35, 37
world demand, 34, 80
cobalt-60, in industrial processes, 63–4
therapy, 61
COCKCROFT, Sir JOHN, 10

COMAR, CYRIL, 79
compensation for radioactive damage, 79
cosmic rays, 22, 84
costs, compensation, 79
 desalination, 57, 58
 nuclear power stations, 35, 38, 39–53
 nuclear waste disposal, 85
 radioisotope power devices, 64
curie, definition, 22

deuterium, as moderator, 15
 in nuclear fusion, 20, 21, 81
 see also heavy water
development, economic, *see* economic development
diagnostic techniques, radioisotope, 60, 61–2

earth movement, 55, 56, 85
economic development, 5, 53, 56–9, 62–3, 64–7, 74, 80–2
EINSTEIN, ALBERT, 11
EISENHOWER, President, 2–3
electric power, 2, 4, 34–5, 56, 57–9, 82, 84
electromagnetic separation, 30
electrons, 8–9, 23
environmental problems of nuclear energy, 5, 72, 73, 79–80, 83–6

fall-out, 22, 55, 77–8, 79, 80, 83–4
fast breeder reactors, 4, 39, 82
FERMI, ENRICO, 13
Fermi reactor, 77
fish contamination, 79
fission, 13–15, 17–20, 28, 54
food, plant mutants, 63
 preservation, 5, 63
 radioactive contamination, 79, 83
France, 24, 25
FRISCH, OTTO ROBERT, 13, 17, 25, 26, 27, 28, 77
Frisch–Peierls Memorandum, 25, 26, 27, 28
fuels, conventional, compared with nuclear, 14
 costs, 44, 45, 47, 48, 49, 51–3
 hazards, 75
 nuclear techniques in extracting, 55, 56, 63–8
 transport, 37

fuels,—*continued*
 waste disposal, 35, 37
 world demand, 34, 80
Fukuryu Maru (Lucky Dragon), 79
fusion, 20–21, 32–3; *see also* thermo-nuclear energy

gamma radiation, 9, 17, 19, 21, 54, 63, 65, 66, 68
gamma-ray detectors, 65, 66, 68
gas, cooling of reactors, 38, 39, 82
 extraction and location, 55, 56, 63, 64, 65, 66
 turbines, costs, 45
gaseous diffusion method of uranium enrichment, 16, 29, 30, 44
genetic effects of radiation, 22, 60, 63, 73, 75, 77–8
Geneva conference, 3
Germany, 2, 44
graphite, as moderator in fission process, 15, 16, 21, 30, 37, 38–9
 for canning nuclear fuel, 39

HAHN, OTTO, 13, 77
half-life, definition, 7
Harwell Atomic Research Establishment, 16
hazards, nuclear, 4, 5, 6, 21, 22, 33, 35–6, 55–6, 60, 72, 73, 74–5, 79–80, 82–6
heart diseases, 5, 62, 64
helium cooling, 38–9
heavy water, 15, 17, 30, 38, 39; *see also* deuterium
HEVESY, GEORGE C. DE, 10
Hiroshima, 1–2, 24, 77, 80
HOFF, HANS, 83
'hydrogen' bomb, *see* weapons, nuclear
hydrogen-3, *see* tritium
hydraulic fracture disposal of wastes, 85

Industrial Revolution, 34, 80
industry, nuclear energy affecting, 5, 21, 34-5, 53, 55–60, 62–8, 74, 80–3, 84
information, *see* public information problems
insects, radioisotopes in controlling, 62–3

international agreement on permitted radiation, 22, 60, 75, 80, 84
International Atomic Energy Agency of the United Nations, 62, 81
international cooperation, 3, 22, 33, 44, 56, 60, 62, 70–1, 72, 73, 75, 76, 80–1, 82, 84
iodine, radioactive, in milk, 83
iodine-131 therapy, 61–2
isotopes, *see* radioisotopes

Japan, nuclear bombing, 1–2
radioactive fallout, 79
joule, definition, 22

Lake Michigan, 77
law and nuclear problems, 70–1, 72; *see also* safety regulations; treaties
light from nuclear power, 5, 64
lithium, 10, 11, 20
location, nuclear power stations, 35–7, 45, 46, 48, 77, 84
peaceful nuclear explosions, 56
Lucky Dragon, 79
luminous paints, 64

Magnox nuclear power stations, 38, 43, 44, 45, 46, 82
Manhattan Project, 25, 29, 31, 77
manpower training, 3, 40, 82
MARSHALL, General GEORGE C., 2
mass number, 10
Maud Committee, 25, 28
medicine, nuclear energy affecting, 5, 21–2, 59–62, 63–4
MEITNER, LISE, 13, 17, 77
mental health aspects of atomic energy, World Health Organization study groups, 76, 78, 83
metallurgy, 59, 63, 67–8
milk, contamination, 83
mining, 5, 55, 63, 64–7, 81
moral aspects of nuclear energy, 1–2, 4, 69–75
MÜLLER, HERMANN J., 78
mutation, *see* genetic effects of radiation
mythology, 76–7, 82

Nagasaki, 1, 24, 77
neptunium, 14, 15, 20

neutrino, 19, 23
neutron, 9–10, 23
neutron activation analysis, 68
neutron detectors, 65
neutron-die-away technique, 65, 67
Non-Proliferation Treaty, 4, 56, 82
nuclear binding energy, 11–12, 14, 17–18
fission, *see* fission
fusion, *see* fusion; thermonuclear energy
Nuclear Power Programme, Second, 82
nuclear wastes, *see* wastes, nuclear

oil, costs, 3, 45, 47, 48
hazards, 75
nuclear techniques in extracting, 55, 56, 64–7
transport, 37
world demand, 34, 80
Operation Sunshine, 80

Pacific, fish contamination, 79
Palomares, 85
PANETH, FRIEDRICH A., 10
Peaceful Uses of Atomic Energy, United Nations Conference, 80, 81
Pearl Harbour, 25, 29
PEIERLS, Sir RUDOLF, 25, 26, 27, 28
pest control, 62–3
pharmaceutical sterilization, 63–4
phosphorus-32 therapy, 61
photosynthesis and radioisotopes, 62
plants, 62, 63
plasma, in fusion process, 21
plutonium, and alpha emission, 15
bomb, 31
fast reactor, 16
fission, 15, 17–18
power source, 20, 64
production process, 2, 14, 16, 19, 20, 27, 30, 31, 37, 38
pollution, environmental, 22, 60, 72, 73, 74–5, 77–8, 79–80, 84, 85
POST, LAURENS van der, 2
prospecting, nuclear techniques, 63, 64–7; *see also* mining
protactinium, 20
proton, 9–10, 23
public information problems, 2, 5, 76–86

radiation, detectors, 65, 66, 68
    effects, on living organisms, 22, 60,
        63, 72, 73–5, 77–8, 79–80, 83, 84;
        on materials, 5, 21, 63
    see also fall-out; gamma radiation;
        international agreement on per-
        mitted radiation
radioactivity, discovery, 7–9
radioisotope X-ray fluorescence tech-
    niques, 66–7, 68
radioisotopes, 10, 13–14, 18–20, 21, 26,
    54, 60
    uses, 5, 21, 59–64, 66–7, 68
radiostrontium, *see* strontium-90
radon, 66
responsibility, moral, 4, 69
Richland, Washington, nuclear waste
    storage, 85
Röntgen, definition, 23
RUTHERFORD, Lord, 9

safety regulations, 4, 33, 35–6, 56, 60,
    73, 74–5, 80, 84
screw-worm control, 62–3
shell structure of atomic nucleus, 8, 10,
    18
sodium cooling, 39
space technology, 5, 32–3, 64
SPORN, PHILIP, 51–2
steel, affected by radiation, 21
    for canning nuclear fuel, 38
    nuclear techniques, in mineral pros-
        pecting, 66; in producing, 59
sterilization of medical goods, 63–4
STRASSMANN, FRITZ, 13, 77
strontium-90, contamination, 79–80, 84
    in power units, 64
    therapy, 61
superstition and nuclear energy, 76–8,
    80, 81, 82, 83, 86

telecommunications, use of radio-
    isotopes, 64
thermal diffusion, 29–30
thermonuclear energy, 3, 4–5, 20–1,
    32–3, 55, 79–80, 81, 84
THOMSON, Sir GEORGE, 25
thorium, isotope production, 19
    location, 65–6
thyroid diseases, 61–2
tracers, radioisotopes as, 10, 60, 61–3,
    66–7

transport, fuel, 35, 37, 53, 81
    radioactive wastes, 85–6
    water, 57
transuranium elements, 13, 14, 19
treaties, 4, 33, 56, 82
tritium, 20, 21, 55, 56, 64
TRUMAN, President, 77
Tube Alloys, 25

underground weapon testing, 33
Union of Soviet Socialist Republics,
    2–3, 24, 56, 81, 83
United Kingdom, Atomic Energy
    Authority, 1
    Radiochemical Centre, 5
United Nations, 62, 80, 81
United States of America, fuel extrac-
    tion, 56
    and Japan, 1, 2
    nuclear accident, 85
    nuclear power costs, 43, 44, 49–50,
        51–2
    nuclear research, 16, 24, 25, 27, 37,
        39, 81
    nuclear wastes, 85–6
    public information problems, 83
uranium, ceramic, 82
    cost, 27, 33, 39, 41–2, 44
    and discovery of radioactivity, 7,
        9–10, 13–15
    enriched, 16, 17, 26, 27–31
    fission, 17–18
    and production of bomb, 25–33
    prospecting, 65–6

vitrification, of nuclear wastes, 85
    in underground explosions, 56

WALTON, ERNEST T. S., 10
War, First World, 33
    moral aspects, 70–1
    nuclear weapons as deterrent to, 4,
        33, 72–3
    Second World, 14, 33, 78, 81
wastes, conventional fuel, 35, 37
    nuclear, 5, 19, 81, 84–6
water, cooling, 35, 36, 37, 38, 56
    desalination, 5, 56–9
    as moderator, 17, 21, 37
    pollution, 55, 56, 79, 84–5
    resources, 5, 35, 36, 55, 57, 58–9, 63,
        65, 80–1
    used in oil extraction, 67

water,—*continued*
  see also heavy water
weapons, conventional, 78
  nuclear, 1–3, 4, 5, 20, 24–33
    conventional explosive equivalent,
      14, 28, 32, 55
    as deterrent, 4, 33, 72–3
    development, 24–33
    moral aspects, 70, 73

weapons,—*continued*
  nuclear,—*continued*
    Palomares incident, 85
    testing, 24, 79
    types, 32–3, 81
    used, 1–2, 24, 77, 80
Windscale, reactors, 38, 83
World Health Organization, 76, 78,
  83